THE CONSPIRATORS

A plausible path of prophetic historicism

An apocalyptic novel by W. R. Reinhardt

The Conspirators

Written by W. R. Reinhardt
Copyright 2022

This novel is a sequel to *Pathway to Destiny,* published August 8, 2022 (eBook) and August 28, 2022 (printed editions). The next novel in the series The Last Apocalypse has been written and awaits publication. .

Unless otherwise stated, all scripture quotations are from *The Clear Word*, a paraphrase Bible written by Jack Blanco.(2006). All rights reserved.

EAN-13 numbers in application.

ISBN 978-1-66787-661-0

ISBN 978-0-9-42442-65-6 *(eBook)*

Book covers (hard cover and Ebook) designed by Self Pub Book Covers/ by R L Sather.

Reader's Review

Reinhardt's novel is a collage of science fiction, philosophy, religion, legend, and mystery. The presentation of the characters and dialogue are such that these novels (**Pathway to Destiny** and **The Conspirators)** would naturally flow into a screen epic. The dialogue and settings are done with clarity. The fact that the dialogue is accomplished without profanity is rare in today's literature. **The Conspirators** is a 5-star work of literature which has the flavor of a screenplay epic.

This book may be presented as one-man's dystopian vision of the future, but much of the story proceeds in line with the drama of the current world's state of affairs. The reason for that many of the events happened as described and reported as news-worthy events. The conclusions derived are therefore beyond mere speculation. When combined with the not-so-subtle political views of the author, the events described lead beyond the theoretical and inevitably to social, religious, political, and military conflict.

This novel is so thought-provoking that I have recommended it to friends, family, and business associates.

Dylan Hoyt

Acknowledgements

There are numerous people to whom I am indebted. I credit Tony Wood who provided a listening ear and constructive criticism night after night. I can never give enough thanks to my sisters Carolyn Moe & Laurie Salmons, my friend Alina Ancheta, and my son Sean Ancheta-Reinhardt who provided technical support and encouragement. They were always willing to listen to my prophetic ravings and my theories of how big problems will be resolved.

About the Author

Due to the environment of the home in which I was raised, and a healthy competition with my older sister Carolyn, I learned to read and write preschool. Some of my earliest memories were bedtime reading of in Maxwell's Bible stories, in which I was encouraged to imagine myself as a key character. By the end of my senior year in high school, at least a quarter of the library's books had my name in the checkout list. I was a superior student and received an academic scholarship to Duke University. In the summer before matriculation, I enrolled in Evelyn Wood's Speed Reading and Comprehension course for six weeks. I graduated from Duke University in 1973.

In the 1980's, I worked for Encyclopedia Britannica in their marketing and distribution. For the next twenty years, I worked in the Securities industry. In retirement, I read hundreds of books of various genre. IngramSpark published my first two novels: *Entry to Alliance Empire* (10/15/21) and *Alliance Regime* (12/1/21). *Pathway to Destiny* was released for publication at the end of August 2022.

LIST OF CHARACTERS

Tom Caine	Vice President of U. S.
Dan Clay	Director of C.I.A.
Sir John Clever	the Terrier
Nigel Cooper	Rothschild's bodyguard
Bill Donovan	Sec. of Homeland Security
Capt. Dustin Garrett	astronaut Lunar Gateway
Colonel Huangdi	Mission chief Cheng'e 6&7
George Ivanovich	Dir. of National Intelligence
Alan Kenney	U.S. Secretary of State
Luis I	the Pope
Lucifer	Lord of the Air
Mephisto	demonic Legionnaire
Gen. Linwood Michaels	SC Orbital Laser Defense
Dr. Ribao	Cheng'e 6&7 physician
David Rothschild	Q of Council of Twelve
Rugby Player*	Council of Twelve Owner/
*	Director of Tesla, SpaceX, and Kilo
Power Gen. Lawrence Thompson	Joint Chiefs of Staff
Morgan Trussler	U. K. Security G45
Dr. Albert Whitehurst	President's Science Advisor
Annette Williams	Pres. of the United States
Captain Charles Wilshire	Avent Security
Yuan Li	demonic Legionnaire

THE CONSPIRATORS

TABLE OF CONTENTS

CHAPTER 1 – Russia Withdraws from International Space Station

Figure 1. The International Space Station in 2018. Source: NASA/EPA

The New York Times (August 7, 2022). The Russo-Ukrainian War's fallout in space includes the International Space Station.

Russia said today that it <u>would withdraw from the International Space Station</u>,[1] which has been a symbol of post-Cold War cooperation between Russia and the U.S. since it <u>was launched in 1998</u>.

The International Space Station (ISS) took 10 years and more than 30 missions to assemble. It is the result of unprecedented scientific and engineering collaboration among five space agencies representing 15 countries. The space station is the size of a football field: a 460-ton, permanently crewed platform orbiting 250 miles above Earth. It is about four times as large as the Russian space station Mir and five times as large as the U.S. Skylab.

Yuri Borisov, the head of the Russian space agency, told President Vladimir Putin during a meeting that Russia would leave the space station after its current commitment expired at the end of 2024.

"Soon we will begin to build the Russian orbital station," Borisov said. "We have selected a desirable orbit and begin to accumulate modules for assembly.

"Good," Putin responded. "The Russian orbital station's primary function will be to conduct research."[2] Borisov had to stifle a chuckle. The true purpose of the planned Russian orbital station was to control cis-lunar space between the Earth and the Moon, and to regulate traffic to mining concessions on the surface of the Moon. If the Russian President had his way, even the collection of solar radiation in the cis-lunar space would require a Russian license.

NASA did not immediately respond to a request for comment. Ned Price, the State Department spokesman, said, "I understand that we were taken by surprise by the public statement that went out," and added that Russia's announcement was "an unfortunate development."

Experts say the announcement dims the prospect of keeping the station operational beyond the end of the decade. In the past, NASA said it intended to continue operating the space station through the end of 2030.

With tensions between Washington and Moscow rising after Russia's invasion of the Ukraine, Russian space officials had made declarations in recent months that Russia was planning to leave.

The first segment of the ISS launches: The Zarya Control Module launched aboard a Russian Proton rocket from Baikonur Cosmodrome, Kazakhstan. Zarya supplied fuel, battery power, and rendezvous and docking capability for Soyuz boosters and space vehicles. Since 2008, Soyuz booster rockets and Russian space vehicles delivered ISS resupplies and have rotated scientists, astronaut/engineers, and other personnel essential to the operation and missions of the ISS. That cooperation is now relegated to history.

The Russian leadership said that their space agency Roscosmos intends to build and operate their own orbiting space station.[3]

Leading U.S. defense contractors discussed Russia's withdrawal from the International Space Station. Conclusions: Withdrawal from the International Space Station by the Russians has major implications for treaties that form the basis of the legal status of cis-lunar space, the moon, and interplanetary space. Exploration and exploitation of resources, including solar radiation, are complicated by travel, exploration, discovery, and subsequent claims by space-faring nations.

NASA and a host of other aerospace contractors have made significant advances in technologies, without which, mankind's progress to the Outer Planets, let alone the stars, would be minimal. Among these contractors are SpaceX, Blue Origin, Northrup Grumman, Boeing Aerospace, and MAJAR Technologies. The Artemis space missions are focused on lunar exploration, the construction of a mammoth solar-electric grid, and the operation of *The Lunar Gateway. The Lunar Gateway* will eventually become the primary refueling station for all NASA missions departing Earth's orbit. Another important function is the role planned for a depot for the accumulation of ice and rare-earth elements.

NASA's long-term goals are even more enterprising. Using the technology and research developed during the Artemis spaceflights, NASA intends to launch future crewed missions to Mars and the stars beyond.

TEST FLIGHT CONDUCTED IN 2022

A diagram shows the path planned for the Artemis I mission to the moon.

*Figure 2 – flight of **Artemis** I*

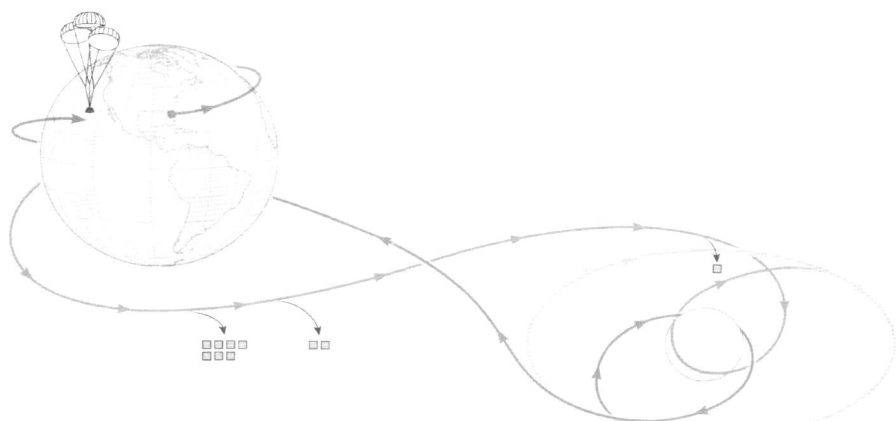

1. *DEPARTURE*
The Orion spacecraft and the Space Launch System rocket will lift off from the Kennedy Space Center in Florida.
2. *MOON ORBIT*
On its way to the moon, the mission will deploy 10 small research satellites called CubeSats. ORION will orbit the moon about 43,000 miles above the moon's surface.
Source: <u>NASA</u> Note: Diagram is not shown to scale.

3. RETURN In preparation for splashdown, the crew module will separate from the service module. The crew capsule will splash down in the Pacific Ocean with the help of parachutes.

MOON
On its way to the moon, the mission will deploy 10 small research satellites called CubeSats.

11

The Orion spacecraft

A diagram showing the different components of the Orion spacecraft.

Figure 3 - **Orion Spacecraft**

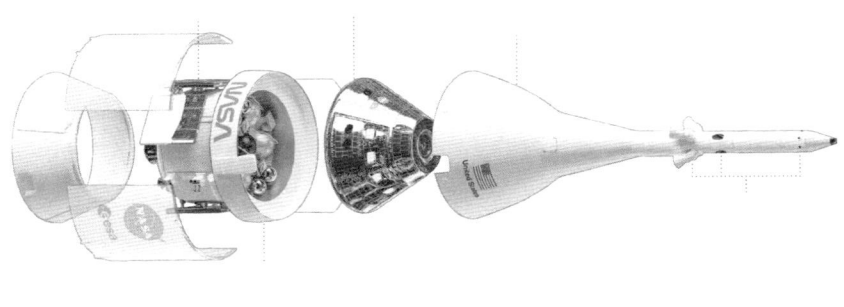

CREW MODULE
Capacity - four people

LAUNCH ABORT SYSTEM
Can carry the crew module to safety if there is an emergency during launch

Solar arrays

Spacecraft
Adapter   Motors

SERVICE MODULE
Provides power and propulsion to the crew module   Source: NASA

Orion will launch into space atop the Space Launch System, a new rocket that stands 322 feet tall and weighs almost six million pounds. The Space Launch System that will be used in Artemis I is one of the most powerful rockets ever developed and can send a payload of almost 60,000 pounds to the moon.

The Space Launch System compared with other rockets

A diagram comparing the height of five rockets: Saturn V, the Space Shuttle, Falcon 9, the Space Launch System and Starship.

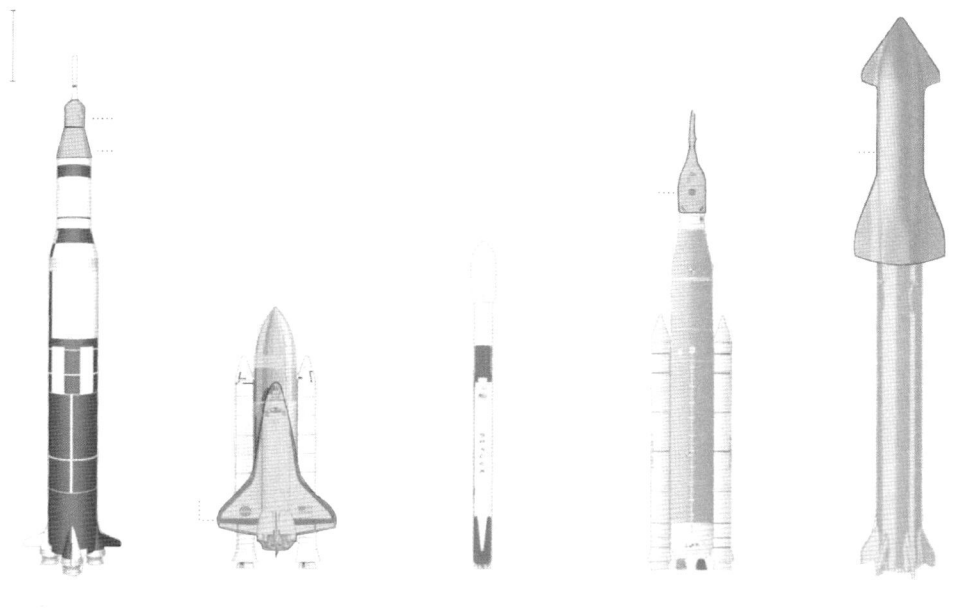

Figure 4. **Past, current, and future rockets**
Apollo command and service modules - Starship spacecraft - Apollo lunar lander
Orion Shuttle
Saturn V - Falcon 9 - Space Launch System
Starship - Space Shuttle
1967—1973   1981—2011

NASA (<u>Saturn V</u>, <u>Space Shuttle</u>)
SpaceX (<u>Space Launch System</u>), (<u>Falcon 9</u>, <u>Starship</u>)

Artemis II

CREWED FLYBY TO BE CONDUCTED IN MAY 2024
A diagram showing the path planned for the Artemis II mission to the moon.

Figure 5

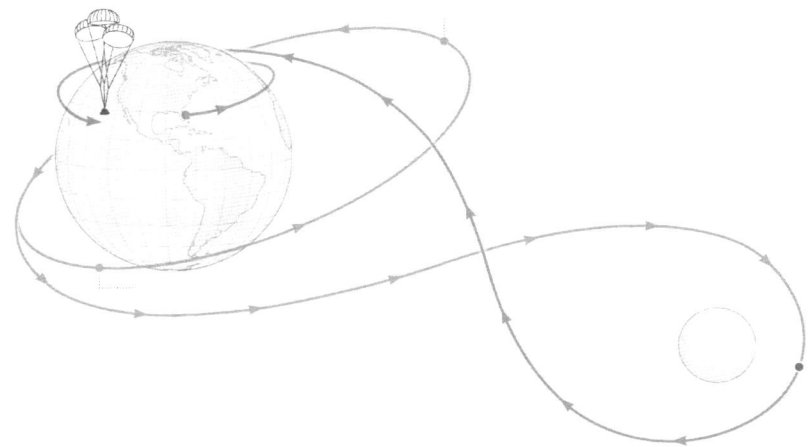

1. DEPARTURE
After launch, the astronauts will orbit the Earth for 24 hours while testing the spacecraft's controls and life support equipment.

2. LUNAR FLYBY
The flight to the moon before the flyby will take four days. *NASA will conduct life support system evaluation and maneuver tests.*

3. RETURN
A recovery boat will collect the astronauts and the crew capsule after they splash down in the Pacific Ocean. Source: <u>NASA</u> Note: Diagram is not shown to scale.

M O O N Flyby
Artemis II, the second mission, will carry four astronauts around the moon and back to Earth. The crew's trip will take them about 4,600 miles above the far side of the moon. The voyage will be man's farthest foray in space.

Artemis III
CREWED LANDING PLANNED FOR 2025
Figure 6. A diagram showing the path planned for the Artemis III mission to the moon.

14

Moon flyby just 70 miles above the surface.

1. DEPARTURE
After launch, the astronauts will enter Earth's orbit, perform systems check, and make solar panel adjustments.

2. MOON ORBIT
While two crew members land on the surface of the moon, Orion and the two remaining astronauts will stay in orbit.

3. MOON LANDING
The lunar lander will undock from Orion and then descend into a lower orbit before landing near the moon's south pole. Astronauts will collect samples.

4. MOON LAUNCH
After a weeklong mission on the surface, the astronauts will ascend into a low lunar orbit before docking to Orion.

5. RETURN
Return the crew to Earth and bring back scientific samples for testing.

In Artemis III, astronauts will land a SpaceX Starship near the moon's south pole while Orion waits in lunar orbit. The polar region is home to mysterious, permanently shadowed craters that have not seen sunlight in thousands of years. The chemicals frozen inside could help scientists understand more about the history of the moon and the solar system.

If the Artemis III mission succeeds, NASA plans to regularly send crews to the moon. Its plans include a lunar base camp and mining operations.

The Chinese have their own plans to build a toll road to the Moon, Mars, and the Asteroid Belt beyond. The Cheng'e Missions are a strong statement of Chinese progress and achievements in cis-lunar exploration and low-gravity engineering.

Ownership rights in space may be determined by an ancient rule called "Might Makes Right." Russia, China, and the United States devised strategies certain to speed up the race to the Moon, accelerate military spending on armaments, perpetuate the current war in Ukraine, and prepare for an imminent war in Taiwan.

The *rivals* in the great controversy are God and his ancient adversary Lucifer. Lucifer strives for domination of the universe within the *Rules of Engagement*. Secret, manipulative fingers within the Council of Twelve exercise great authority in the European and United States Aerospace industries. What innovative technologies can

be developed? How can these technologies assist mankind establish a presence in the outer planets of the solar system and their civilization among the stars?

What armed conflicts have been instigated, controlled, or heightened to the point of nuclear war? Perpetual war is a strategy perpetrated by Lucifer to coax man's most evil deeds. Vladimir Putin, Joe Biden, and Xi Jin-Ping alike fear to take steps, however, that lead to Mutual Assured Destruction.

What natural disasters, such as earthquakes, or strikes by asteroids, will God permit? Will viral pandemics wipe out mankind? Will storms and, alternatively, drought, continue to become worse before mankind deals effectively with global climate change? Will mankind continue to permit the poisoning of Earth's oceans and the atmosphere with *greenhouse gases* until the Second Coming of Christ?

Will God permit sinful man to carry disobedience and other evil behavior to infect the unfallen creatures of God's creation? The Final Apocalypse mark the years that precede the destruction of the Earth. "The Lord will come as unexpectedly as a thief in the night. When He comes, the sky will roll back with a roar, and everything on Earth will be set on fire. Even the basic materials that compose the Earth will melt under this intense heat. The whole Earth will be ablaze and everything on it."[4]

The Good News is that God's gift of salvation is bestowed on those who love Him and seek to accomplish His will in their lives. Mankind is called upon to live a holy life. "Therefore, control your thoughts, stay alert, and be ready for action. Set your hope totally on what Jesus Christ has done for you and what He will do for you when He returns."[5]

CHAPTER 2 - Lunar Gateway

Figure 2. *The Lunar Gateway* in 2024.

NASA and its corporate allies built a space station in lunar orbit. Certainly, anything more than a fuel depot would seem to be a tremendous extravagance. A fuel depot orbiting the Moon at a LaGrange Point is critical. A spacecraft leaving Earth must overcome a gravitational pull that consumes 75% of the rocket fuel carried at departure from the surface of the planet. Also, one could argue convincingly that the functions of the Lunar Gateway and the Kennedy Space Center make the expensive Artemis boosters and the Orion space capsule reusable. *The Lunar Gateway* reduces the booster fuel required for the voyage to Mars since the gravity of the Moon is a small fraction of Earth's gravity, so two of three boosters of fuel are unnecessary and can be saved for reuse or fuel storage. *The Lunar Gateway* reduces the time required for a voyage (one way) of over 140 million miles. The Orion spacecraft departing the Moon refuels at *the Lunar Gateway* and intercepts the planet Mars in its orbit around the Sun without requiring multiple *sling-shots* (gravity assists). Sling-shot assistance is time consuming; the time required for the trip to Mars *without refueling* would increase by nine months.

"The purpose of the *Gateway* is really as a waystation to get to the surface of the Moon and push on to Mars," said the director of global sales and marketing for Boeing's aerospace division.

The company's director, who is *the Candlestick* in the Council of Twelve, could be described as the architect of *Lunar Gateway*. While working at NASA around 2011, he was tasked with figuring out what was next in human exploration of space. "I sat down and went through every single design reference mission NASA had produced for the last 20 years. The files were over 900 gigabytes," he said. "After months of assimilating this information, I reported to a committee for the purpose of prioritizing NASA's goals in space. I took that knowledge and experience to Boeing and began to dialogue with leadership of other aerospace corporations. Before long, I became an integral member of the Council of Twelve."

"I believe that our activities have contributed to the efforts to decarbonize and utilize resources that otherwise would have been wasted. Without the Council of Twelve, space exploration could be cancelled or at least would be years behind current progress."

"*The Lunar Gateway* has opened a lucrative market for solar energy which previously had been served by fossil fuels. Dependence on crude oil, natural gas, and coal has poisoned the atmosphere of the planet and acidified the oceans. As many as 30% of all species of life no longer exist. That is what I call unwarranted extravagance."

"Who knows what the cost has been to mankind whose health has been impaired from drinking polluted water and breathing polluted air? Greenhouse gases have elevated planetary temperatures which have caused undesired climate changes and fierce storms. Zero carbon emission is an achievement that is in the realm of possibility. I take immense pride in the global mission of the Council of Twelve," said *the Candlestick*.

CHAPTER 3 – Perpetual War in Ukraine

- Figure 3. Aug. 24, 2022. Ukrainian Army artillery unit in action

WASHINGTON — President Biden said on Wednesday that the United States would deliver nearly $3 billion worth of arms and equipment to Ukraine, its largest single package of military aid aimed at helping the nation battle Russian forces.

The announcement, on Ukraine's Independence Day and the six-month anniversary of the war, signaled Mr. Biden's continuing determination to assist in the fight against Russia's invasion.

In a statement, Mr. Biden said the latest financial assistance would allow Ukraine to purchase "air defense systems, artillery systems and munitions, counter-unmanned aerial systems, and radars to ensure it can continue to defend itself over the long term."[6]

The effort to bolster Ukraine's military, which has garnered bipartisan support in Congress, has now delivered more than $10 billion worth of weapons and other equipment. The aid announced on Wednesday is part of the $40 billion assistance package Congress approved in May.

Mr. Biden acknowledged the suffering of the Ukrainian people and pledged to ensure that the country's sovereignty would be protected.

"Thousands have been killed or wounded, millions have been displaced from their homes, and so many others have fallen victim to Russian atrocities and attacks," he said. "But six months of relentless attacks have only strengthened Ukrainians' pride in themselves, in their country, and in their 31 years of independence."[6]

Since Russia invaded Ukraine in February, the White House has prioritized sending weapons and military goods from the Pentagon's own stockpiles to Ukraine, authorizing more than $8 billion in rockets, missiles, firearms, vehicles, and other hardware from Defense Department supplies.

But Wednesday's announcement could signal a significant shift in how the United States will support Ukrainian forces in the future. The $3 billion President Biden pledged will go to the Ukraine Security Assistance Initiative, a fund that allows Ukrainian leaders to purchase military goods directly from the defense industry.[7]

Shifting the source of Ukrainian military supplies from the Pentagon's own stockpile, which is large but not limitless, to items newly manufactured by the defense industry indicates that the White House and military leaders are transitioning to a sustainable model Kyiv can depend on for an open-ended war with Russia.

"This announcement represents the beginning of a contracting process to provide additional priority capabilities to Ukraine in the mid- and long-term to ensure Ukraine can continue to defend itself as an independent, sovereign and prosperous state," Brig. Gen. Pat Ryder, a Pentagon spokesman, said in a statement emailed to reporters. "It is the biggest tranche of security assistance for Ukraine to date."[8]

The fund was set up in the wake of Russia's illegal annexation of Crimea in 2014, and according to a report from the Congressional Research Service, disbursed more than $1.3 billion to Kyiv from 2016 to 2021, which has purchased sniper rifles, rocket-propelled grenade launchers, counter-artillery radars, patrol boats, night vision devices and other matériel.

According to General Ryder, the $3 billion announced on Wednesday will be used for six new NASAMS air-defense missile systems and additional ammunition for the two NASAMS. launchers the United States previously provided with Security Assistance Initiative funds in July, as well as up to 245,000 additional rounds of 155-millimeter artillery ammunition and 65,000 rounds of 120-millimeter mortar ammunition — up from the 561,000 155-millimeter and 20,000 120-millimeter rounds previously sent to Ukraine. The money will also purchase more drones and anti-drone systems, laser-guided rockets and radars designed to track incoming enemy artillery fire.[9]

Michael D. Shear is a veteran White House correspondent and two-time Pulitzer Prize winner. John Ismay is a Pentagon correspondent in the Washington bureau and a former Navy explosive ordnance disposal officer. Photo is courtesy of *The New York Times*.

Figure 4 – NASAMS Air Defense System

CHAPTER 4 – **Hope of Mankind**

Unless the builder of the house is the Lord, the workers toil in vain. "[10]

London, United Kingdom

David Rothschild possessed inherited wealth from a venerable banking family. David, well-known as a financier, is involved in global enterprises, a philanthropist, and a collector of antiquities. Privy to an elite few, is David's secret identity. David is 'Q,' the organizer and chairman of the Council of Twelve. 'Q' has extensive international business and social connections.

Rothschild had attended the upper-class Ravenscroft prep school where he rubbed elbows with the sons of the scions of industry and government. He went on to graduate at Eton. His schoolmasters and professors made an indelible impression. "Honor and Responsibility" was the motto inscribed on the stone column entrance to Eton. David Rothschild learned to value and endorse ethical considerations. His decisions were not heavily influenced by his Jewish heritage; Rothschild is not religious. He considers himself to be a materialist if his thoughts in that regard were to be labeled. Rothschild's ethos was forged in a technical philosophy of materialism, a valid belief that that he did not think could be discredited. His philosophy held that nothing exists but natural phenomena. Rothschild's materialism is a code "of life and living according to rational processes with intellectual and other capacities of the individual to be developed to the highest degree in a social system where this may be possible . . . There were no supernatural forces, no supernatural entities such as gods, or heaven, or hell, or life after death."[11] There were no supernatural events, nor could there be. Rothschild thought

God's existence to be rationally improvable, and therefore at best a meaningless proposition. What proof existed that the Big Bang at the origin of the universe had a supernatural cause? Rothschild thought that all formal religions were forms of superstition. A notion that forces of Light and darkness such as Christ and Lucifer could be contending in an invisible arena for the fate of the world is absurd to Rothschild. The only judgment that Rothschild feared was the record of history.

Rothschild is the descendent of several generations of English bankers, all of whom were well placed in British society. Thus, to a considerable extent, Rothschild inherited a network of connections that were social, industrial, and in the realm of finance. These connections are extensive. His family, colleagues, and business associates considered David to be a gentleman of intellect and unusual competence. As David's career progressed, he became known to military and government leaders.

David possesses a form of quiet courage, and a self-confidence born of planning and discipline. He takes pride in his record of taking business strategies from inception to successful outcomes. David never intentionally caused harm to man nor beast, and he despised cruelty and insensitivity. David contributed considerable time and resources to certain charities.

Rothschild had never sought leadership, per se. His competence and ability to analyze and deduce made him a natural candidate for positions of authority in business. In banking circles, he is known to have the Midas touch. Rothschild preferred anonymity and avoided publicity.

His decision two years ago to organize the Council of Twelve was not driven by megalomania, but by desperation. David believes that mankind was headed pell-mell for extinction and that those in control of the earth's people and resources were doing precious little to prevent disaster. Just as nature abhors a vacuum, 'the socio-economic organism' demands special leadership when there is a vacuum of such leadership. David believed that he was simply stepping into that vacuum, by reaching accords with like-minded individuals in the Council. Iron-clad resolve and combined resources would bring about the events that were necessary to provide a future on Earth for humanity. The Council of Twelve would succeed in critical matters wherein predecessor New World Order organizations such as the Club of Rome and the Committee of Three Hundred had failed. Rothschild considered the ethics of potential actions of the Council of Twelve. In his mission to save mankind, Rothschild chose to bear responsibility for the altered destinies of hundreds of millions, if not billions of human beings.

Recruiting the individual members of the Council was undertaken in the same fashion as any of Rothschild's long-term investment strategies – with caution and careful planning. Visits and communiques were carefully orchestrated and completely confidential. Council members kept no secrets from 'Q'.

The initial entry fees of $10,000,000 each for the twelve positions in the Council were subscribed; one potential member who was tardy in submitting the entry fee was quickly replaced. In the sixty days that had transpired since the Council positions were subscribed, several of the members privately recruited financial partners within their positions.

The members of the Council represented a cross-section of the wealthiest and most influential people on the planet. Five of the members of the Council have large equity positions as major contractors in *Lunar Gateway*. Their purpose of organization was united – plan an effective strategy to deal with impending global catastrophes. The global catastrophes complicated not only mankind's entry into space, but jeopardized mankind's survival on Earth as well. The overriding imperative was to ensure their own survival and at least some portion of mankind. Retaining certain NASA contractors, such as SpaceX, Northrup Grumman, and Maxar Technologies, was critical. Key decisions were made more often in the boardrooms of these defense-oriented corporations, and then approved by NASA.

The identities of the Council were secret. Secure private communications were of utmost concern for the first video conference.

Figure 5. **The Council of Twelve convened.**

With preparedness complete, Rothschild called for a conference.

"But in vain are men's dreams of progress, in vain all efforts for the uplifting of humanity, if they neglect the One source of hope and help for the fallen race."[12]

CHAPTER 5 - **Council of Twelve**

London, United Kingdom

March 23 12:00PM

Member Icons

the Terrier   the Oil Derrick   the Crossed Sabers   the Top Hat
the Automobile   the Cargo Ship   the Pyramid   the Hammer
the Rugby Player   the Candlestick   the Hourglass   the Falcon

Members of the Council displayed only an icon when logging into the video conference. The use of an icon was one method to protect the confidentiality of their identities. The voices of the Council members were modified by a voice synthesizer whenever they spoke. Now, one by one, the icons of the Council members appeared on the individual monitors on the table in the conference room.

'Q' convened the Council without ceremony. As the meeting commenced, 'Q' was seated and alone in the conference room. As the meeting progressed, council members and staff came and left as duties necessitated. A keyboard at his fingertips lay on the desk in front of him.

He rose with arm extended in a straight salute, reminiscent of the Third Reich. Yet David was not a Nazi.

"I pledge allegiance to the Council of Twelve. I pledge to hold its missions and programs in secrecy. *I resolve to defend the human race and the planet Earth from all threats within and without."*

"A quorum of members has been attained at 12PM Greenwich." Whenever a member of the Council was speaking, the icon which represented that member flashed. The icon that 'Q' had chosen was a light gray skull superimposed on a Doomsday clock. Unless the icon of the speaking member was minimized, the icon of that Council member dominated the monitor.

"This conference is now in session. The Council members will adhere to Roberts Rules of Order. The agenda of each meeting will always summarize the actions or decisions of the previous meeting, then review use of funds, current programs, and then consider new business. Use your 'enter' key or double-click on your icon to speak. The computer program will recognize speakers in order. The program will not permit more than one speaker at a time."

"From initial receipts of $120,000,000 – the following expenses were incurred." Expenses were displayed. "The next business is global threat assessment. For purposes of prioritizing threats to human survival, a range of mid-case to worst-case scenarios have been forecast by our scientists. The threats that I now display are the worst, ranked in order."

The Council members had many interests in common. Most members owned or had monetary interests in companies which denied or minimized the danger of global climate change. These members even employed 'company' scientists – meteorologists, seismologists, marine biologists, agricultural chemists – who produced and issued reports to influence publications and public perception.

But privately the individuals of the Council knew and recognized the realities of the danger of mass extinctions. *Privately, these members had conceded to David Rothschild that **the combination of climate change, over-population, perpetual warfare, the breakdown of society, and viral epidemic could overwhelm mankind.***

"I present threat assessment reports which offer a clear picture of current conditions. *This assessment is confidential.* The reports are available today for a proprietary format. The Council must agree that there will be no unauthorized access to these reports. Any attempt to copy, print or share this video will result in sanctions."

'Q' began the video presentation, entitled 'Confidential Report of the Emergency Planning Committee.' The narration accompanying the ensuing video was deep and somber, much like the voice of famous orator Earl Nightingale:

"The absence of a world war thus far in the 21st century could be a short-lived anomaly. The era of the Cold War following World War II was marked by post-colonialism, nationalist insurgencies, and terrorism. In recent years, technology, economics, and the battles for scarce resources have dominated global politics. The era which is ending, *Pax Americana*, is a misnomer characterized by perpetual limited warfare which is defined by one superpower, two rivals, and their allies while the world nervously endures the possibility of Mutual Assured Destruction. MAD is a short-lived albeit catastrophic conflict in which hundreds of nuclear weapons target all the planet's population centers and military bases. As a strategy, MAD is a deterrent of insane proportions in which occurs an irrational push-button exchange of Inter-Continental Ballistic Missiles. Such a war would render the earth uninhabitable. Over time, public fears about MAD have waned, yet military nuclear strike capabilities of the *nuclear club* - U.S., Russia, France, China, North Korea, Great Britain, Israel, India, and Pakistan – remain. Nuclear disarmament and SALT II are in limbo."

"Tensions between the current great military powers of the United States, Russia, and China rise precipitously. Increasingly, the leadership of those countries choose hardline approaches, favoring confrontation over negotiation."

"Conflict with Russia grows as Russian drones and submarines constantly probe the borders of the NATO member states. Current Russian military adventures in Ukraine including the seizure of Crimea press NATO to a high state of alert. Former Iron Curtain/Warsaw Bloc nations are mobilized to respond to Russian military aggression. Economic sanctions are imposed which are tantamount to a blockade of Russian goods and services. The war between Russia and Ukraine is perpetual, with no end in sight, giving NATO and its allies many opportunities to test theie latest weaponry."

"Turmoil with China intensifies. Chinese economic and military power provide evidence of China's legitimacy as a superpower. The Chinese bargain, cajole, and demand trade concessions and sovereignty over contested territories. Increased border incidents and worldwide cyber-hacking can be attributed to the Chinese military-industrial complex. These activities provide strong signals of eminent military conflict. Chinese military defense planners gave insight to the strategic thinking in their PLA Journals of National Defense with this quote: "China must consider the possibility of a third world war when developing responses to provocation."[14] The most volatile issue is Taiwan.

"The technologies add new levels of complexity and burgeoning costs to the arms race. Although offensive military capabilities for satellites and other space-based weaponry are banned, both the U.S. and

China spend billions on several types of satellite and anti-satellite weapons. *Jane's Journal of Armaments said that it was an even bet as to whether China or the U.S. would first deploy their illegal weapons in space.*[15] More billions are spent in the development and production of a new generation of fighter-bombers; the superpowers have unmanned jetfighters and bombers nearing production. These aircraft augment the tens of thousands of drones already fielded. New warships capable of firing electromagnetic rail projectiles have been unveiled, superseding eight hundred years' use of gun powder. Radical changes in the design and composition of warships and submarines offer substantial automation with greater firepower and stealth. For infantry, technological advances in personal weapons and armaments make the modern soldier more formidable than ever. Rumors persist of Russian scientific breakthroughs in artificial blood, resident artificial intelligence, and computerized exoskeletons for the world's first trans-human warriors."

"Each superpower progresses in its strategies for world domination. China is a late entry in the race for space but has a well-funded space exploration program and has developed effective anti-satellite weaponry. The Chinese army experts in cyber warfare and espionage are matched only by the best of the West. Chinese training and education far exceed U.S. efforts in the field of cyber mischief. At risk are U.S. and European civil and industrial infrastructure, commerce, and the entire array of communications, both civilian and military. Chinese armed forces are modern, massive, and completely prepared for conventional warfare. Deep caverns serve as last resort refuges for the Chinese military-industrial elite. Russia as well maintains a conventional war machine prepared to exploit any weakness of NATO in Europe. Russia has rapid strike capacity with the world's second largest submarine force, half of which lurks just outside U.S. territorial waters."

Figure 6. *Russian subs lurk in U.S. territory.* (news.usni.org)

"Equally disturbing is the recent adoption of a first-strike strategy by U.S. joint staff planners. Widespread bombardment of electromagnetic pulse could fry electronic devices, controls, and digital communications in Russia and/or China. Thus U.S. foes could either be paralyzed or sent back to the pre-electronic era. The U.S. and allies currently maintain obsolete equipment, including jet aircraft, ships, weapons, and non-communication technology from the pre-electronic period, in the event of countermeasures by the Chinese or Russians. Bulletins and procedures are distributed for civil preparedness, reminiscent of the Cuban missile crisis of 1962. Certain industries providing essential services are being placed in wartime footing to achieve a pre-electronic era redundancy. Industry is racing to lay underground and protected cable. Electric grids and power plants are making what many consider urgent preparations for war."

"Many violent conflicts rage worldwide. There are civil wars in four Islamic countries; some of these struggles threaten to split and divide these countries along sectarian lines. Even more troubling is the sabotage of a nuclear reactor in Iran; Israel is widely suspected. Both nations are exchanging threats and mobilizing their armed forces for war. Iran is threatening shipping in the Strait of Hormuth. A regional war seems inevitable, a conflict which could widen. The U.S. Navy has fleets in support positions in the eastern Mediterranean and the Red Sea. The Russians have countered by building up its Baltic Sea fleet and conducting joint military exercises with their allies."

"Global climate change has accelerated." The video presented a series of clips which showed scenes of drought and wildfires, torrential storms and flooding, and rapidly melting ice from the polar regions. Renowned meteorologists presented the case for strengthened phenomena, such as El Niño and polar vortex. Data is shared regarding the increasing accumulation of greenhouse gases in the atmosphere and the oceans. Evidence is provided that undersea volcanic activity is warming the world's oceans to alarming levels unprecedented in modern history.

"Greenhouse gases absorbed into oceans and lakes are increasing the acidity of these large bodies of water. Slowly but surely, earth's water is becoming inhospitable to life."

"Plankton, protozoa, and tiny mollusks are threatened; since these are the food sources for most of the larger fish and sea mammals, the fishing industry is doomed."[16]

"The jellyfish, already found in massive quantities in the oceans, will share the oceans of a dying planet with vast beds of anaerobic red algae."[16] A list of extinct species scrolled through the film. "Jellyfish play important roles in the marine ecosystem and are a key source of food for some fish and sea turtles.

Some even protect commercially valuable species, such as oysters, from predators."

"Some species of jellyfish are suitable for human consumption and are used as a source of food and as an ingredient in various dishes. Edible jellyfish is a seafood that is harvested and consumed in several East and Southeast Asian countries, and in some Asian countries it is considered to be a delicacy. However, the ascendency of jellyfish is the demise of other species. For scarce biotic resources, jellyfish are the ultimate competitor."

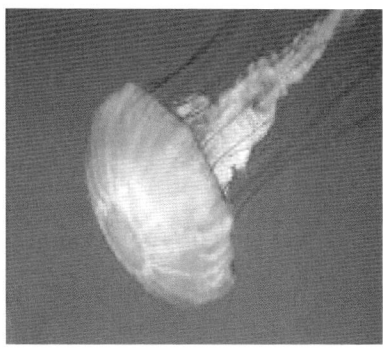

Figure 7 – Jellyfish proliferate in oceans.

Figure 8. ***Food Aid for storm and drought victims*** *(wfpusa.org)*

The video segment on the environment ended dramatically with scenes of the emaciated bodies of starving children without a prospect of their next meal.

'Q' continued the video presentation without interruption. These giants of industry and commerce viewing the presentation from the comfort of their corporate offices held their own glum assessments and were willing to wait for the general discussion.

"Civil society is rapidly disintegrating," the narrator proclaimed, "and is in imminent danger of collapse. Anarchy, terrorism, gun violence, suicides, human trafficking, drug trafficking and drug addiction – these activities have reached unconceivable levels."

The video depicted scenes of urban and rural warfare as economies and societies collapse in Latin America, South America, Africa, and the island nations. In the industrialized countries, particularly the United States, an increasing dependence on opioids rendered tens of thousands of people dysfunctional. For others, the drug of choice was heroin or meth-amphetamines. Public Education was in free fall. The traditional family was a vanishing institution.

The final segment of the video presentation highlighted the threat posed to humanity by disease. "The likelihood of pandemic disease becomes greater and greater. SARS viruses strike viciously, then mutate into milder strains or variants. Similarly, short-lived outbreaks of Ebola are reported in the basins of the Congo and Amazon rivers. In the United States, the Center for Disease Control has declared an emergency; over ten million Americans have developed infections which are resistant to antibiotic medication. Many infections are asymptomatic. Productivity losses by year end are expected to be more than three hundred billion dollars. Worldwide, the opportunistic diseases of cholera and diphtheria accompany the flooding and starvation that follows big storms. Deaths from cancer, heart disease and conditions related to diabetes continue to climb. Outdoor air pollution, specifically PM 2.5 pollution, decreases the average life span of a human on the planet more than road injuries, HIV-AIDS, malaria, and war combined." The narration was supported by graphs of statistics.

'Q' ended the video and opened the discussion. "I propose that humanity's suffering is exacerbated by *over-population.*"

"The problems that over-population create have compounded the difficulties man faces with natural causes of global changes. Over-population puts pressure on scarce resources and geopolitical boundaries. Over-population breeds anarchy and terrorism. Over-population provides an opportunity for global pandemics. The rain forests of Africa and South America which have historically produced vast quantities of oxygen have been decimated to feed this burgeoning population. The capacity of the agricultural and commercial fishing industries to feed this growing population is strained to the limit. In the process of making all this fertilizer, vast amounts of fossil fuels are consumed, which exacerbates our global carbon dioxide problems. In summary, no one can doubt that the efforts to sustain the daily activities of eight billion people have produced dangerous levels of greenhouse gases. The human race increases by millions each year. According to a United Nations Census report last year, an estimated 160 million were born and sixty-five million died, which resulted in a net increase of ninety-five million people on the planet. Yet no government in recent years has dared to curb in their country the run-away growth of population."

"At this time, we will consider recommendations of the Council."

CHAPTER 6 - **Debatable Solution**

The icon of *Crossed Sabers* flashed. "Have you created a model which predicts the combined effects of these threats to global population?"

"Yes. I will summarize," said 'Q.' "*The combination of disasters expected to befall the planet could reduce the population by ninety percent - as many as seven billion will die over the next thirty years. Without our planning and resolute action, the marine life in the oceans will die, the planet will experience hellacious extremes in weather with both flooding and drought, agriculture will fail, the global economy will collapse, and nuclear war is likely.*"

Several icons began flashing.

The Rugby Player spoke. "Any combination of these disasters obviously disrupts our financial empires. Can a radical plan be implemented to reduce or alleviate these impending crises?"

'Q' responded. "Our research teams have prepared a video presentation which includes a proposal for a post-holocaust civilization with a planned economy. The global population would be capped at five hundred million. Our scientists calculate that with zero carbon emissions, much of the global climate changes can be reversed. A reduced population would not inhabit storm-threatened coastal areas. Land use would be restricted from the development of interior land threatened by persistent wildfires or repeated cycles of tornados. Areas prone to earthquake or storm damage will not be settled without improved construction standards."

The Oil Derrick asked, "Do you have a comprehensive plan to achieve zero carbon emissions?"

'Q' answered, "Our consultants in the energy industry plan to present a detailed proposal based upon the estimated requirements of a drastically reduced world population. This proposal is based upon consideration of existing technologies. I expect some assistance from Council members in this effort." 'Q' paused. "At this point there are way too many variables to suit me."

The *Terrier* was incensed. "The most principal factor is the reduction of humanity's excess population!"

'Q' considered this opinion. "Let us see if our Council members have consensus on this matter, as this issue impacts strategy and methodology. The *Terrier* proposes that *population reduction is critical to humanity's short-term survival and moves the Council to implement such strategies necessary to reduce global population to five hundred million.* Is that correct?"

The *Terrier* delayed his response for a few seconds, then spoke firmly. "I so move."

"Is there a second?" 'Q' inquired.

The *Pyramid* was first to respond. "Second."

'Q' proceeded. "Is there any discussion?"

The *Top Hat* was next recognized. "When the population of a species exceeds the capacity of the habitat to sustain that species, the population of that species must be 'culled' in the interest of the survival of the habitat." The statement provided moral justification for radical action.

One by one, most of the entities represented by the icons expressed similar sentiments.

Then the *Candlestick* spoke. "My associates and I control the largest agri-business conglomerates, which include the manufacture of industrial fertilizer. You mentioned that your expert consultants had considered 'existing technologies.' I have information which has a bearing on both the over-population and the achievement of large-scale carbon sequestration."

The Hammer interposed. "Do your scientists believe that we can achieve zero carbon emissions *and* reduce accumulated greenhouse gases?"

The Candlestick responded. "Yes."

The Top Hat pursued this query. "I have no confidence that our reliance on fossil fuels *can be reversed.* Increased population and their activities will generate more carbon emissions."

The Candlestick said "My studied opinion is that under ideal conditions, the planet could sustain a very much larger population, but unfortunately, our governments have waited too late to create those ideal conditions. The Director of this Council correctly identifies one of the myriad problems which beset our industry. I will not subject the Council to an enumeration of all those problems, but the issue that the Director raises – the immense quantity of fossil fuels used in the manufacture of fertilizer – is worthy of a brief discussion of case in point."

"There is a plant – a tiny fern – that has a unique ability to capture carbon dioxide and nitrogen from the atmosphere. This fern is called Azolla."

"Since the dawn of agriculture, subsistence farmers in southeast Asia have deliberately cultivated Azolla as a companion plant for rice.

Azolla is a floating fern that thrives in rice paddies, fixing nitrogen and other nutrients, constantly improving the soil composition, and providing a natural green fertilizer that significantly bolsters rice productivity."

The *Candlestick* continued. "The secret here is that Azolla is not just a plant; it is a 'superorganism,' a symbiotic collaboration of a plant and a powerful microbe.[16] In a special protective cavity inside each leaf of this tiny fern, Azolla hosts a microbe called Nostoc. Nostoc spends its entire life converting atmospheric nitrogen into food for its host. I am talking about a plant that can double its entire body mass in less than two days. Azolla and Nostoc have clearly demonstrated an immense ability to combat global warming and to convert atmospheric nitrogen into compounds that could help feed the world."[17]

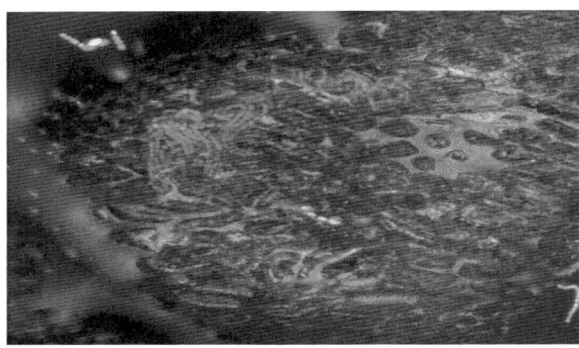

Figure 9. Super-organism Azolla with Nostoc (figshare.com)

"Recognizing an opportunity, one of our companies invested over $10,000,000 for plant genome studies to unlock the knowledge embodied by this superorganism. Our research has revealed much of the language that codes for the molecular machinery underlying this symbiotic partnership. We have been able to design genetically modified variants of these organisms tailored to our needs for the major grains, fruits, and legumes. These GMOs have been identified and patented."

The *Crossed Sabers* queried, "When will these products be marketed?"

The *Candlestick* responded. "We had planned to license Monsanto to produce and market these GMOs. But those plans are on the shelf."

"Why?" interposed the *Top Hat*.

"There are two major reasons: Water and politics. The use of Azolla/Nostoc GMOs will revolutionize current farming practices. The methods of planting, nurturing, and harvesting cash crops as 'companions' to Azolla are radically different. A great deal of water suitable for agricultural use is required. *The success of Azolla farming would eliminate most of the fertilizers in use today.* That is a reduction of over $100 billion for global consumption of nitrogen fertilizer. Markets for the farming equipment and the pesticide industries will be reduced. However, there is the potential for significantly increased crop yields. Gone also is the pollution of streams, rivers, and estuaries from the 'runoff' of fertilizer. Our industry made proposals to the Senate committee on Agriculture and the U.S. Department of Agriculture. We made similar proposals at the Summit of G-7 Nations at Copenhagen."

Again, from the *Top Hat*. "What happened?"

"Nothing," said the *Candlestick*. "In addition to the price supports and the low return on investment for farmers, there are too many vested interests to overcome. The requirement for more water in Azolla farming is a complicating factor. Fertilizer requires a lot of petroleum. The fact is that our proposals are mired down in politics."

The *Oil Derrick* spoke. "Our group represents most of the world's petrochemical industry. Regarding 'existing technologies,' I also have information which has a bearing on both the over-population and the achievement of large-scale carbon sequestration. First let me state that there is no simple or easy answer for reduction of greenhouse gases, recovery of the delicate balances of ecosystems, or the reversal of de-speciation. Unbelievably, the energy industry has always been receptive to 'going green.' There are many promising technologies. We could harness the power of the wind to a much greater extent. Our industry could invest heavily in solar power; we could invest in satellites to collect and retransmit microwave energy. Our industry has revolutionized battery technology for bikes and motor vehicles. Our group owns patents on competitive technologies which can slash carbon dioxide emissions and generate vast quantities of oxygen. The problem has always been that the politicians do not want to pay the price. The failure of leadership has led to our current crisis."

The icon of the *Automobile* spoke. "Can you provide an example of one of the 'competitive technologies' that removes carbon dioxide and generates oxygen?"

The *Oil Derrick* responded. "Certainly. One of our companies purchased the patents for 'engineered cyanobacteria' that uses sunlight in a photosynthetic process to produce N-butanol and/or pentanol. In the process of producing a gallon of butanol, the 'designed' bacteria remove about sixteen pounds of carbon dioxide from the atmosphere and give off about sixteen pounds of oxygen. If that gallon of butanol ends up being burned in an internal combustion engine, such as an automobile, the exhaust would release the sixteen pounds of carbon dioxide back into the atmosphere. Although this operation is a zero-net gain, the outcome is vastly improved when N-butanol is substituted for additives in gasoline such as ethanol. N-butanol has about 50% more energy per gallon than ethanol and burns at a lower temperature. Auto manufacturers have great interest in N-butanol."

The *Oil Derrick* continued. "We constructed a pilot plant in Essen, Germany, which was completed in early January last year. Our 'phytonics' facility produced over twenty-five million gallons of N-butanol; all of which was sold to the industrial chemical market. The plant produces N-butanol in volume for about $1.50 per gallon, whereas other production methods cost over $3.00 per gallon. Yet ethanol, which is converted from corn in the United States, gets the big government subsidy and tax credits from the politicians, while elsewhere in the world, people starve. Go figure that one out."

Most of the icons were now illuminated; this was a subject of much interest.

The *Oil Derrick* had more to say. "Allow me to defuse suspicion that our petrochemical industry is in conspiracy to withhold 'magic bullets' or that we have been withholding immediate solutions to resolve complicated problems that have been around for a long time. Bear in mind that such an immediate solution for the problem of carbon emissions, resulting from the global demand for cheap fuel, does not exist. And if such a solution did exist, market forces would destroy the global economy. Our industry has oil and natural gas reserves estimated at over $14 trillion with refineries and other infrastructure worth another $1.5 trillion. These assets would be devalued, and millions of people would be unemployed."

"Let me share our experience with our 'phytonics' facility. Surely this manufacturer producing N-butanol and pentanol is no secret to the city of Essen, or the 265 Germans employed there. But the plans to construct four more production facilities are on hold."

The icon of a *Cargo Ship* flashed. "Why?"

Oil Derrick: "Overtures were made to the automobile industry, which has retooled significantly for the electric-hybrid market."

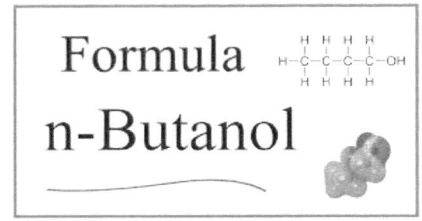

Figure 10 – symbiotic collaboration of plant and microbe

https://www.google.com/url   molecular structure for n-Butanol

"The automotive industry would have to retool again, since N-butanol has a much higher octane. The engines and fuel injectors would have to be modified. Similar overtures were made to the airline industry. Pentanol is an outstanding aviation fuel. But Boeing and other aircraft manufacturers have not converted their jets for the use of pentanol, despite its many advantages. Again, retooling is the primary issue. There is also the matter of tariffs and taxation. These industries would have invested the capital to retool had there been intelligent government leadership. In summary, the petrochemical industry needs economic incentives to develop our technologies and market our alternative fuel products. But political leadership is fractured, divisive and corrupt."

The Falcon icon was recognized. "We need strong leadership."

The Terrier interposed. "Or more likely, we need to control social forces, manipulate the media, select own our own leaders, and ruthlessly promote our agenda!"

The Pyramid commented. "Bravo! Spoken like Machiavelli."

The Crossed Sabers spoke. "I have no interest in listening to any more hand-wringing over the matter of over-population. The Council has convened to make some tough decisions. *The elimination of excess population* is one of them. Let us vote."

'Q' closed discussion and put the motion to a vote. One by one, 'Q' made the roll call. *The Terrier's* motion passed unanimously.

The *Rugby Player* posed a question. "You mentioned a 'post-holocaust' phase in your plan for humanity's survival? When would this phase begin?"

'Q' responded. "The report predicts circa the year 2038."

The Hammer flashed. "Does your research team have a back-up plan, say, a failsafe alternative?"

"Yes," 'Q' said. "Thanks to *the Rugby Player*, *the Candlestick*, *the Top Hat*, and other Members, we have realistic plans to colonize outer space and amass vast fortunes."

The conference was abuzz.

The *Terrier* was online in Skype with three other Council members during the conference.

At the conclusion of the electronic meeting of the Council, the *Hammer* said, "Rothschild is a sanctimonious fool. We have the majority among the membership to seize control and coerce the agenda. Why do we wait?"

"Be patient. Rothschild is a useful tool. He will bring to the table many opportunities for profit. Wait until the Russian and Chinese oligarchs are recruited. Then we shall deal with Rothschild when the time is right."

CHAPTER 7 - **Manipulated**

Kingdom of the Air

March 23

From the dark matter netherworld, Lucifer and Mephisto witnessed the meeting of the Council of Twelve. *The existence of the Council and, to an extent, its initial organizational success is manipulated by the demon Legionnaire Mephisto.*

Mephisto was concerned. He knew that the *Terrier* was secretly a Luciferian, a worshiper of Lucifer. Yet the business of the Council had an unexpected outcome to which Lucifer had expressed satisfaction. The demon and his master had watched the proceedings of the Council. As the invisible viewing portal faded away, Mephisto had questions.

"Sire, I do not understand. *Overpopulation* has long been our objective. As a population outgrows its resources, the result historically has been warfare, disease, and starvation, with the best and brightest of the humans dying first," said Mephisto. "But the Council has elected to radically curtail Earth's population, solve the planet's environmental problems, and create a utopia where everybody is one big happy family!"

A deep profundity of laughter burst out as Lucifer and his minions responded. As the barrel roll grew silent, Lucifer said, "You act like the buffoon. Must I retain your services as a court jester for comic relief? The best laid plans of mice and men so often go astray. Your job is to make these plans collapse in a spectacular fashion. You must *guide their lofty aspirations of materialism into debauchery.*"

Instantly, Mephisto and Lucifer teleported to an underground vault stacked with currency. Various signs were posted on the shelves in German.

"Where are we?"

"Deutchebanc vaults," replied Lucifer.

"Whose currency is stored here?" Mephisto was mystified.

"Only the currency of the United States in denominations of $20s and $100s," said Lucifer.

"How much cash is available at this location?"

*Figure 11 – **Billions stored in U.S. currency***

"Billions and billions."

"In the age of digital transactions, why is all this cash necessary?"

"These funds are mostly the proceeds of human trafficking and the drug trade. After laundering, the funds go to legitimate businesses. Digital transactions are traceable. The ubiquitous dollar is not. Hence the U.S. dollar is the medium of choice for international criminal enterprises," said Lucifer.

Lucifer then transported Mephisto from bank vault to bank vault in city after city in every continent.

Mephisto saw billions in U.S. currency held to launder illicit trade.

The last scene that Lucifer and Mephisto viewed together was a riotous party like out of *Animal House*. Businessmen in suits were gathered at tables while naked women performed for them. At the center of the establishment, heat radiated from a grill about ten' square. Low flames licked at the red-hot grill. Loud throbbing music and strobe lights provided a magical atmosphere.

"Who are these people?" asked Mephisto. "It looks like a frat party."

"Commercial money market traders. They make the entire system function," said Lucifer.

"What is the large flaming grill for?"

"Human sacrifices, mostly children. Luciferians really know how to please me." Lucifer's demented staff hooted with glee.

Lucifer Accuses God ~ 1900 B.C.

"Representatives from various planets in God's created universe come together regularly to meet. . . One day Lucifer also came and presented himself for admittance. . .God said to Lucifer, "On what basis do you want to attend this meeting?" The accuser answered, "I represent Earth. I am in charge there and have been for a long time."[18]

CHAPTER 8 - **Rules of Engagement**

London, United Kingdom

June 19   12:00PM

Council Member Icons

the Terrier   the Oil Derrick   the Crossed Sabers   the Top Hat

the Automobile   the Cargo Ship   the Pyramid   the Hammer

the Rugby Player   the Candlestick   the Hourglass   the Falcon

'Q' convened the Council of Twelve. "A quorum of members is attained at 12PM Greenwich. The Council is in session." The icons on Q's monitor were all back-lit. 'Q' quickly reviewed actions taken at the first Council session and current expenses. At the Council meeting, updated and modified threat assessments and analysis reports were presented by video.

After the previous meeting, encrypted reports had been sent for the subtle purpose of seeing which Council members would adhere to protocol. A security trap had been laid and sprung. Several Council members had ignored the security protocol required; attempts were made to print, save, or copy reports. Attempts for any activity other than read-only tripped programmed security loops, and the offending members were shut out of Council communiqués. Payments of significant fines re-instated these members. Recording now was less likely. The dissemination of Council reports is a danger that all Council members understood, and a risk that none would knowingly undertake. 'Hacking' and clandestine eavesdropping were guarded against by talented industry specialists. Nasty surprises are in store for even the most experienced hacker.

The members awaited the video report which proposed survival and recovery plans from worldwide pandemic, perpetual war/ mutual assured destruction, and global climate changes that would destroy all life on earth. The video report proposed alternatives – failsafe operations that include radical expenditures to curb pollution, impose universal disarmament, invest in immunological research, and extend the boundaries of human life to the outer planets.

'Q' advised the members that the presentation would require three hours, with short intermissions.

The reports were a sophisticated production consisting of video clips of recent catastrophic disasters, along with confidential information from the United Nations Security Council, the National Security Agency, and the Center for Disease Control. Interviews were conducted with the world's leading environmental scientists, meteorologists, and seismologists. The highlights of the past year's earthquakes, volcanic activities, hurricanes, cyclones, tornados, wildfires, and flooding were viewed.

These clips were followed by the warfare and armed conflicts raging in Africa, the Middle East, Indonesia, and the Philippines. There was a segment on the drug cartel wars in Central and South America. Private opinions were elicited from former secretaries of state from several countries.

The Council members heard a report and analysis of global societal breakdown. Gun violence, particularly in the United States, a marked global increase in suicides, and growing addiction to drugs and alcohol provided somber realities. Actuaries described causes for the reduction in life expectancy; the year 2022 marked the fourth such reduction in succession. The reductions in life expectancy corresponded with coronary-vascular disease, poisonings from environmental pollutions, cancer, diabetes, and other forms of degenerative disease. *A secret report from the CDC warned of anticipated pandemics.*

The risk of nuclear holocaust had increased due to proliferation of nuclear devices and advances in miniaturizations. Additionally, the United States and Russia had failed to implement Strategic Arms Limitation Treaty reductions in nuclear weaponry. Russia and the U.S. had recently recalled their ambassadors. Détente seemed unlikely any time soon. The current risk of a nuclear war was greater than at any time since the Cuban missile crisis of 1962. Posturing by North Korea and threats of nuclear attacks persisted. India and Pakistan were locked in a face-off of armies in Kashmir. Mutual Assured Destruction would not deter these nations from self-destruction. Computer simulations displayed graphically the results of a limited exchange of nuclear weapons, and then a simulation of an all-out war with military bases and municipalities struck by Intercontinental Ballistic Missiles. A limited exchange would accelerate a nuclear doomsday; a war with more than 150 ICBM strikes would exterminate mankind and most other species of life.

The report concluded with graphs of predicted increases in greenhouse gas, with the associated effect on storms, drought, and increased acidification of the oceans. A doomsday clock appeared at the end of the video production; the time shown was twelve minutes until midnight.

'Q' spoke. "The Council is open for motions and discussion."

The Terrier icon flashed, and a deep simulated voice was heard. "The planet is quickly becoming an inhospitable place to live. My associates and I control five multi-national conglomerates. We believe that over-population exacerbates the doomsday risk. This crushing burden can be alleviated. One of our companies is a leader in pharmaceutical research and immunology. *We have designed viruses that can inflict high mortality and sterilize most of the world's males*. The laboratory has also created a vaccine to immunize our own children. Research and development on the debilitating virus have proceeded for some time and testing of prototypes is underway. This research has been kept, until now, completely contained to a few scientists. My associates and I require 'assurances' before we make any attempt to utilize our research."

The Rugby Player icon flashed now. "What do you call this virus?"

The Terrier responded, "The host virus is SARS-CoVID-2. The virus carries a modified human papilloma virus, which causes the sterility. The condition is called 'the Crimp,' which is a characterization of the blocking effect that the modified virus has on the path of the sperm from the gonads. We have drafted a plan to effectively disseminate the virus and its host on a global scale."

'Q' spoke. "Excellent. Presuming that the Council can provide adequate assurances for *the Terrier,* does any Member have a motion?"

The icon of *the Pyramid* now flashed. "I move that the Council provide such assurances. This information must never be disseminated to any unauthorized person."

'Q' moderated. "Since each Member is already pledged to secrecy regarding the business of the Council, what further assurances does *the Terrier* propose?"

The Terrier spoke. "Verbal assurances are meaningless in these circumstances. I understand that a few Members have already violated security protocols for secrecy of Council reports. Therefore, I propose to put some 'teeth' in the commitment of the Members to Council business and for severe sanction in the event of loose lips or whistle blowing. The cost of implementing the sterilization program is $120 million. I move that each member agree to an assessment of $10 million for this purpose. My associates and I are ready to act upon a precise plan immediately. The Council can hold in escrow the Members' assessments until the World Health Organization affirms the existence of a new viral epidemic. At that time, the Chairman would release our funds."

The Hourglass flashed. "I second *the Terrier's* motion subject to an appropriate discussion. I have a master's degree in biology, and I have a few questions about the virus and its transmission."

'Q' spoke. "The second from *the Hourglass* is accepted. The Council will now entertain the motion from *the Terrier*."

All the Icons were backlit for commentary and questions. The computer recognized *the Candlestick*. "I can appreciate the need for valid 'assurances' from each member. Monetary assessments will accomplish a great deal more than blood oaths or dark threats. I will presume, unless other Council members are forthcoming, that no one else is prepared to deal with the dilemma of overpopulation in as effective a manner as a deadly pandemic which leaves survivors sterile. I await further details."

The Falcon spoke next. "*Terrier* – what else can you tell us? My associates and I are intrigued, but we press you for more details. We will not provide such an *assessment* on a whim."

The Terrier responded. "I can offer further details on the program for human sterilization. But prior to the Council's receipt of your assessments, I will not provide such details which would pinpoint certain laboratories, or the research conducted in those laboratories. The scientists and lab technicians are bound to secrecy. In the industry, our group is known by reputation for our advanced research in the field of human fertilization. You may be aware of a recently approved prescription medication which is marketed by the trade name 'Activate.' This drug chemically induces a minute electro-magnetic charge which assists in achieving contact and union between ova and sperm. To obtain the reverse or opposite effect, an electro-magnetic charge creates a blockage at the juncture of the vas deferens and the seminal vesicles. Our research scientists call this effect 'crimping;' thus, the condition, which causes male sterility, we have labeled 'the Crimp.' Our labs have designed and produced a virus which is an engineered variant of HPV. A highly contagious low-grade SARS acts as host for the HPV. The SARS virus is of short duration, but the HPV which is transmitted persists and is resistant to almost any medical treatment. Our lab has also prepared a vaccine for the virus."

The Terrier paused and continued. "My associates have retained an independent contractor who will recruit young men in the 17-21 age range to participate in what they will be told are clinical trials of a new medication. These 'clinics' will then expose the young men to our virus in eighty of the world's largest metropolitan centers. The clinics will only operate for eight weeks at each location. These young men will spread the viral contagion and transmit the variant HPV."

The Rugby Player interjected. "The authorities will trace the epidemic back to the Council members!"

The Terrier responded. "The short duration of the operation of the clinics and the natural incubation period of the host virus will give our contractor time to close up shop and disappear. Plus, we have purchased 'insurance' that the business of the clinics will never be disclosed. Our clinics will be opened and closed long before the local health authorities investigate. You should know that these clinics will mostly operate in 'third world' countries."

The computer recognized *the Automobile*. "Do you intend to achieve a random distribution of this virus?"

The Terrier answered. "No. These clinics will operate in Africa, Southeast Asia, the Philippines, Indonesia, Central and South America, and the megacities of China. Although the virus will spread world-wide, the greatest effect of the virus will be achieved in eighty metropolitan centers in those areas."

"Mr. Chairman." *The Terrier* addressed 'Q.' "Will you please present the computer simulation for the spread of the 'Crimp' contagion?"

The monitors of the Members all displayed an enlarged icon of *the Terrier* for a few moments, then a slide show image of two hemispheres of the globe was shown. *The Terrier* moderated. "In this next slide, a bright red glow represents each of the metropolitan centers selected for the

'Crimp' clinics. Thereafter, the slides progress in thirty-day intervals for nine months."

The Council members watched in silence as the spread of the 'Crimp' progressed in red from onset to month nine. At first, the 'Crimp' seemed confined to an area surrounding the metropolitan centers at origin. The viral contagion spread faster and faster. By the seventh month, fully half the planet glowed red, and showed some effect in Europe and North America. The final slide – month nine – the whole globe was red, with the brightest glow marking 'third world' countries.

All the Icons flashed. *The Top Hat* was the first to be heard. "How soon could the clinics be open for business?"

The Terrier spoke. "The contractor is on standby, awaiting the outcome of our deliberations, and the receipt of the necessary assessments. Twenty clinics will be opened at a time, with the first group opening on August 1st. The last group of twenty clinics will end operations on March 30th of the following year. All arrangements, including rental sites for the clinics, advertisement copy, equipment, supplies and technicians, are ready. My associates have even identified the local officials who need to be bribed."

The Hourglass spoke. "The transmission rates of different virus vary widely. What are the assumptions of your slide show presentation?"

"Our virologists have designed a SARS virus that we expect will exceed the highest transmissibility of any virus ever seen."

"Bear in mind that this particular virus causes a low-grade fever, and that most individuals who 'catch' the virus will suffer a few days of nausea and diarrhea, as well as runny sinuses."

"As always with respiratory tract illness, the virus may present a major complication for an individual who has a weakened immuno-response. Infants and the elderly will be at higher risk and have higher mortality. Many will die. We expect that the World Health Organization and the Center for Disease Control will designate this virus to be an epidemic. Remember that the host virus – SARS – does not cause the sterility. The variant HPV causes the sterility. Virtually everyone who contracts the variant SARS is infected with the HPV. The course of the SARS virus will terminate. But the variant HPV will continue to spread through sexual contact. There are anticipated side effects for women."

The Hourglass spoke. "What are these side effects for women?"

The Terrier answered. "Women who contract the variant HPV will have a markedly increased risk of cervical cancer."

The Hammer observed, "The selection of the target population centers is discriminatory!" The voice paused. "I like it."

The Hourglass: "What consideration has been given to the potential fallout? Are there unintended consequences to our actions? For example, the Chinese will not fail to perceive the lopsided distribution of this mass sterilization. Will they go to war?"

The Hammer: "To hell with the communist Chinese! None of the SARS-Covid clinics targeted Taiwan, Japan, or the Koreans. One can rationally conclude that their own bio-hazard laboratories, like the one in Wuhan, had an accidental release of the virus."

The Hourglass: "I have concern about sabotage – the communist Chinese are formidable adversaries. Until our defenses are in place, our new space station is a big fat sitting duck. Eventually, the Chinese need to be recruited and included in the business of the Council."

The Terrier: "The Chinese do not 'play well' with others. They will obstruct the critical activities of the Council. At this point, the Chinese would not be good partners in our endeavors. Once the sterilization is a *fait accompli*, the Council can deal from a position of strength with the Chinese, and the Russians or other potentially worrisome groups. The Council could "consider" an application from a 'qualified member' after the virus has achieved its purpose. Is that a workable scenario, Mr. Chairman?"

'*Q*' spoke. "Yes." *Very shrewd, Rothschild thought. I wonder. . .*

The Pyramid spoke. "What about the antidote inoculations?"

The Terrier responded. "The antidote inoculations are my biggest concern for the possible breach of security. We will assign paramedic technicians to inoculate those individuals that you select. I will suggest a procedural protocol and provide inoculations on a reasonable schedule. How you induce your family, friends, or employees to receive their inoculations is your business, but the actual reason for these inoculations cannot be divulged!"

The Crossed Sabers spoke. "This is the action step required by our earlier decision to reduce the planet's excess population! I cannot conceive of an alternative plan which produces the desired effect with less damage to the planet. The logic of this plan to de-populate and sterilize most survivors is unassailable. If we can drastically reduce new births, then the predicted natural and environmental disasters, combined with casualties from war and other causes of death, will achieve our desired population cap."

     '*Q*' thought '*Why do I suspect that the Terrier is ready to implement this program with or without the participation of the Council? Clearly other members have made similar conclusions.*'

     The Rugby Player proposed, "Let us close discussion and vote."

     '*Q*' spoke. "Assessments must be received by June 30th. If there are no objections, I will close discussion of *the Terrier's* motion, and conduct a roll call vote. Due to the nature of this motion, a unanimous decision is required, with no abstentions."

     There was no quibbling over the assessments, and the vote was unanimous.

     The business of the Council concluded an hour later. The Council discussed the initial funding of *the Lunar Gateway*. The estimate for the construction of a space station orbiting the Earth and the Moon at Lagrange point four was $1.7 trillion. *The Lunar Gateway* was heavily subsidized by NASA, the European Space Agency, and a Who's Who list of western aerospace industry.

     The contractors controlled by Council of Twelve Members are SpaceX, Boeing, Northrup Grumman, Blue Origin, and MAXAR.

     The *Lunar Gateway* is an expensive proposition, but its budget was nowhere near NASA's proposed Deep Space Gateway – the most expensive project in the history of man's era in space. Significant resistance came from business and the public to spending upwards of two percent of the nation's Gross Domestic Product on the Deep Space

Program for the space expedition to Mars. The NASA plan requires assembly of modules that would be delivered to orbits mostly ranging from 1,000 to 43,000 miles above the lunar surface. A Space Launch System would deliver the large components yet to be built. The final assemblies prior to the Mars Expedition would be made in Earth-lunar orbit at LaGrange Point 4, mostly by robots and self-assembling components. *Lunar Gateway* would be much less expensive, and revenues would be forthcoming more quickly.

Lunar Gateway is the private venture of proxy partners of the Council of Twelve. Public interest is remarkably high. *Lunar Gateway* would be a failsafe refuge for those fleeing a dying planet and a waystation to Mars and the stars in the very possible event of the postponement or cancellation of NASA's Deep Space program. Currently the Deep Space program would launch in the year 2030. Regardless of whether the NASA Deep Space program ever becomes operational, then *Lunar Gateway* would be open for business.

Figure 12 – design of *NASA's proposed Deep Space Gateway*

"To be controlled by human nature results in death; to be controlled by the Spirit results in life and peace. A person becomes an

enemy of God when he is controlled by his human nature; for he does not obey God's law, and in fact he cannot obey it. Those who obey their human nature cannot please God." [19]

CHAPTER 9 - **New Imperative**

Kingdom of the Air

June 19

The image of Lucifer appeared in Mephisto's mind. Every other perception – sight, hearing, and even the sense of feeling – fled from his awareness. When Lucifer wanted the attention of one of his demons, he permitted no kind of distraction.

"Legionnaire," said Lucifer in his booming baritone, "Do you question my strategies for the Council of Twelve?"

Mephisto was immediately alarmed. He knew that humans who have given themselves over to addictive behavior and satanic delusion had surrendered their minds for easy manipulation. Mephisto and Lucifer had communicated mind-to-mind for over six millennia. Could Lucifer, without invitation, read Mephisto's mind? Mephisto did not doubt it for a second. The demon was glad that he had shared none of his doubts with the minions of his legion.

"Sire, I do not question the brilliance of your strategies, but yes, I do confess that I do not understand how your strategies will advance our global agenda. Could you enlighten me?"

"Regarding the sterilization program or the orbiting space colony?" Lucifer queried.

"I have a handle on the space colony. Our rebellion has bragging rights to the universe of created beings that our dominion has expanded beyond the planet," allowed Mephisto. "We create a secular society on the space colony that denies or minimizes the existence of God. Since only the wealthy and the technocratic elite can afford to live in the colony, it becomes a man-made utopia. With reduced gravity and longevity therapy, life spans for individuals could be centuries. Most importantly, all the credit for these developments goes to *man*, not God."

"Outstanding," Lucifer glowed. "What are your misgivings about the massive sterilization?"

"I always thought that overpopulation was our strong suit in this deck of cards. Too many people on the planet grubbing for increasingly scarce resources. The outcome of their struggles is frequently bad – starvation, genocide, poverty, *perpetual warfare*, lack of educational opportunity. The many disasters that befall mankind provide lots of opportunity to mischaracterize God in specific and Christianity in general." Mephisto presented his case.

"All important considerations, my dear Legionnaire, that were important in the past," Lucifer confided.

"In the past?" echoed Mephisto. "What happened?

"From our knowledge of Biblical prophecy, we remain certain that the next events in our Adversary's timetable are the loud cry of the Three Angels' Message and the second Pentecost, which the remnant church calls the Latter Rain.

'Yes, yes,' thought Mephisto. Every demon knows these things, and this knowledge brings fear and trembling. All the denizens of the Kingdom of the Air are aware that time is short, and that after the Adversary's return to Earth, there will be 1,000 years of confinement for the fallen angels on a desolate and depopulated Earth. Then will come the final judgment. The future for Lucifer and his legions of demons becomes increasingly desperate.

"We now have a New Imperative," announced Lucifer. "Our demons no longer need to lead mankind down the primrose path to destruction. Mankind is doing an adequate job on that score all by themselves. So, we must help mankind – *in particular, our allies against God* - avoid extermination. The atmosphere and oceans of the world are past the point of recovery. Draconian measures, including mass sterilization, will only delay the inevitable. Mankind is doomed. The battle henceforth is entirely a spiritual battle for the minds of men and women. Deceit, delusion, and disinformation will continue to be the tools of our trade, but we must moderate to an extent the destruction that mankind has brought upon themselves."

"Remember that the whole universe of God's created beings bear witness to this controversy. How will the death throes of the planet Earth appear to the unfallen beings of the other worlds? I predict that after the last man dies or leaves the planet, we will be able to arbitrate our fate at the Conference of the Worlds. Therefore, the final judgment becomes hypothetical."

"Your job is to enable the Council of Twelve to achieve their goal of a population cap. I employ many other capable demons who will *deceive mankind into believing that they can save themselves.*"

The Dragon Thrown Out of Heaven

"(7) *The controversy between God and the dragon began years ago in heaven. God's Son Michael and the loyal angels fought against the dragon and his angels.* (8) *The dragon and his angels fought back but were defeated and lost their place in heaven.* (9) *The great dragon, called the devil and Satan, that ancient serpent who is leading the entire world astray, was thrown out of heaven and came down to the earth with his angels.* (10) *. . . Salvation and the power and the kingdom of our God have come, and so has the authority of Christ to rule. The accuser of our brothers who accused them day and night before God and has been defeated and is cast out of heaven forever.* (11) *Those who he accused overcame him by their faith in the blood of the Lamb and by their personal witness, and they did not shrink from giving their lives in death for Him who loved them.* (12) *Rejoice you angels and you who dwell in the heavens, but woe to the earth and the sea, because the devil is extremely angry and is coming at you with great fury. He is filled with rage because he knows his time is short.* (13) *When Satan saw that God had been vindicated, he pursued the woman who had given birth to the male Child.* **Revelation 12:7-13.**

CHAPTER 10 – **Demons Confer**

Kingdom of the Air

October 25

"...the devil is extremely angry and is coming with great fury. He is filled with rage because he knows that his time is short."[20]

Mephisto occupied an exalted position in Lucifer's top hierarchy of fallen angels. As such, he was privy to inner council and accomplice to many of his master's end-time strategies. Of late, Mephisto lobbied ceaselessly for more resources to conduct his role in the attack on the Remnant church, specifically "those who keep the commandments of God and hold to the testimony of Jesus Christ."[1]

Mephisto's Legion commanded one hundred Centurions, who led commands of one hundred demons, ten thousand in all.

As time marches relentlessly to the spiritual battle of Armageddon, Lucifer's leading commanders direct the activities of over one billion demons. Principally these demons focus their energies on distracting men and women from the important matters of salvation. There are many diversions. Chief among them is materialism; sham religion and self-indulgence follow suit.

The rank-and-file demon simply observes human behavior and deduces what sort of temptation would be most effective. Demons know that repetition leads to addictive behavior. Often the relationship between the demon and the human's thought pattern becomes so close that the demon can project a thought or desire into the human's mind. The ultimate relationship occurs in "possession" wherein human beings become hosts for demons and their activities.

The resources for which the Legionnaires compete are the services of the most effective fallen angels. The acquired knowledge and skills of unimpaired demons are highly prized by Lucifer. Exacerbating the shortage of highly competent demons is an ever-increasing incidence of *melancholy disease* afflicting the fallen angels. Many of them suffer acute depression, others had simply gone mad. After more than six millennia of separation from their Creator, as many as one-fifth of the fallen angels suffer some form of impairment. devils, Lucifer had to imprison in stasis the very worst of the devils gone mad.

Figure 13 – Demon gone mad, imprisoned in stasis

Although the angels had been designed to exist forever, now once-perfect features were twisted and stunted by the ravages of aging. When Lucifer complained to God about the melancholy disease and the aging process, God merely said, "My creatures were made to bask in My presence." In desperation, to keep devils from fighting Mephisto influenced the activities of Illuminati groups like the Club of Rome and the Committee of three hundred since the inception of the 20th century.

Demons are heavily invested in the Council of Twelve. Many fallen angels under the direction of Mephisto's Centurions silently assisted in the dissemination of the modified influenza virus. Some of the world's largest cities are under a special Satanic assault. The Council of Twelve are using young men as volunteers for a medical test. The clinics operated in each city only as long as the virus serum took to incubate in their human hosts.

A tone chimed in Mephisto's brain. A telepathic conference was signaled with the Master.

As usual, several of Lucifer's aides-de-camp listened into the exchange.

"All allegiance, Sire." Mephisto's response was proforma.

"Duly noted, Legionnaire," said Lucifer.

That voice, thought Mephisto, the mix of bass, tenor and baritone was a relic from the ancient past – from Lucifer's duties as the covering cherub for Jehovah God. The tone of the master's voice could change with his mood. There were some voice tones that Mephisto hoped never to hear. Mephisto served Lucifer with equal parts of respect and fear.

"Do you confirm that our agents are in place at the clinics in each city?"

"Yes, Sire. Please note that some of the clinics are not yet open for business. Our lab investigators have determined that the viral antigens as designed by the pharmaceutical firm will produce a swine-linked influenza. The World Health Organization will designate the strain of the influenza, and their labs will find that our strain is resistant to antibiotics, antiviral drugs, and the current CDC inoculation. The targeted rates of infection predicted by the Council of Twelve will be achieved."

"And the variant human papilloma virus?" Lucifer inquired.

"It is too early to expect results from the field yet for the HPV. But the transmission method works, and the projected rates of infection are reasonable. Our agents in the public domain will entice many young humans to engage in the sexual encounters that will spread the disease. After the HPV propagates, we will promote a disinformation campaign by the chief officials in public health. Governments, particularly in east Asian countries and the Third World, will deny that the decline in birth rate persists. Our demons will convince the officials to believe a lie until it is too late to mount an effective response. The Council of Twelve will accomplish their goal for radical reduction in global birthrates, unless..." Mephisto hesitated.

"Unless what, Legionnaire?" Lucifer's voice had a new and ominous edge. Mephisto knew better than to mention that there might be interference from God. The greater the danger that a virulent contagion posed to mankind, the more likely it became that God would step in to alleviate the threat. *God might invoke the Rules of Engagement.*

"Unless the Council of Twelve self-destructs." Quick thinking could sometimes rescue a demon from Lucifer's wrath. "The individual members of the Council have giant egos and their own private armies. As we speak, members conspire against each other."

Mephisto heard a humming in the inner-most part of his mind while Lucifer conferred with his aides. Mephisto waited respectfully.

Lucifer spoke, "Remember to instigate murder and mayhem, but not at the expense of compromising our major strategies. The sterilization must proceed. It is critical to my plans. So far, so good." Coming from

74

Lucifer, this remark was high praise. "I trust that among your Centurions there is one who can direct the tasks of the Council of Twelve."

Mephisto heard, or sensed, a muffled grunt and soft laughter in the part of his cortex where the telepathic meeting took place. It was annoying to be outside of the commentary and the snide remarks of the demons on Lucifer's staff. Undoubtedly the Master tolerated the rude behavior by the aides-de-camp for reasons that Mephisto could not appreciate. Mephisto became instantly wary. What did Lucifer imply? Did the Master intend to burden Mephisto with yet another responsibility? Worse yet, did Lucifer intend to promote him to his staff? Curses! A demon's work is never done. Mephisto was exceedingly confident of successfully managing his current assignments, but he recognized his own limitations. Did Lucifer realize that Mephisto and his Centurions were constantly at work? Or was there some other problem? Were Lucifer's aides jealous of Mephisto's many successes? Did Mephisto's old foe Yuan Li want him to fail in his current endeavors? Mephisto knew that he had no friends on Lucifer's staff. Jealousy was an age-old tradition of the Kingdom of the Air.

Mephisto worded his response cautiously. "My command objectives include the Council of Twelve, the Papacy, a secret USAF space weapons program, and several other special projects that you yourself have imposed. All assignments are at a critical stage."

"Allow me to specifically address the SARS epidemic that the Council of Twelve has contrived. For the last five hundred years, I have trained thousands of demons in the sciences of epidemiology – virology, immunology, pathology, and laboratory technique. Many of the demons that I supervise are veteran virologists of famous pandemics such as the bubonic plagues of the Middle Ages, Ebola outbreaks in Africa and devastating attacks of influenza. Whenever medical research is conducted, I have placed demons for the purpose of perverting progress and designing yet another debilitating disease." Satan cast a stern glare at Mephisto.

"Legionnaire!" Lucifer declared. "Your work will not be taken from you!" The timbre of the master's voice echoed in Mephisto's head. "No one else is as qualified. Listen to what you must do." The aides-de-camp were eerily silent.

Lucifer continued, "I want you to map the gene sequences for the 1918 influenza virus to reproduce it. Design a flu strain that originates in chickens or swine that infect humans. Produce a virus that the human immune system has never experienced. I want a virus that strikes young adults hardest."

Mephisto was disappointed. The Master had forgotten Mephisto's greatest achievements in virology as well as some recent history. In 1918, Mephisto had produced the most virulent strain of influenza ever seen. The pandemic swept the world in three waves. The second wave was most destructive. It triggered an immune response so great that white blood cells attacked every organ in the human body. Later many medical researchers referred to Mephisto's accomplishments as "the cytokine explosion."

Mephisto knew better than to bring attention to Lucifer's apparent memory lapse. Reminding the Master would require some tact. Better not dismiss a bad idea out of hand, especially if the idea originated with Lucifer. This was an opportunity to score some points for his team and silence the enemies he had acquired on Lucifer's staff.

76

"Jeff Taubenberger and other researchers at the Armed Forces Institute of Pathology reconstructed the entire 1918 virus and extracted its genomic sequence in 2005.[21] Since then, human scientists have learned a great deal about how my demons have adapted flu viruses from chickens and pigs to humans. My team has taken advantage of poor sanitation, polluted water, and filthy, crowded animal slaughterhouses to create ideal conditions for spreading disease. Thus, there are multiple opportunities for the viruses to jump from animals to human hosts. One of the biggest breakthroughs was when we learned how to tweak eighteen subtypes of hemagglutinin and eleven subtypes of neuraminidase so that a virus can mutate from strain to strain."

Mephisto heard an audible groan from Lucifer's staff, but he persevered. "My command group has an unmatched record of success in producing killer influenza pandemics. Since 1510 A.D., we lay claim to causing over a million deaths on seven separate occasions. Most of these humans died cursing God with their last breath! And in my crowning achievement, from 1918 to 1919, more people succumbed to influenza than were killed in all the battles of World War I!" Mephisto felt a satisfying murmur of assent from Lucifer's staff.

"Do you have a counter proposal?" asked Lucifer.

"I need not remind you that time is of the essence. What do you foresee as the mortality rate for the new pandemic?" asked the Devil.

"My Centurions wager that one quarter of the global population will suffer illness in the first wave of the pandemic, and ten percent of those infected will die. If your other legionaries can cause warfare and starvation, then the virus will be opportunistic. We have engineered a mutation which brings about a second, more virulent wave of infection. Then we may well see a mortality rate of 25% of those affected."

"Bear in mind," said Mephisto, "that there is potentially an even greater motivation for this mischief."

"Indeed," Lucifer affirmed. "a radical reduction in the birthrate of humans to match their reduction in lifespan." The aides-de-camp hooted and stomped their feet. A high-pitched laughter was cut off as the ethereal connection ended, leaving a hollow silence.

Mephisto was shaken by the impromptu conference, which had done little to boost his confidence in Lucifer's leadership and satanic end-time strategies. Questions remained unsolved in Mephisto's mind. What use could the Master make of a plunge in the human birthrate? Or for the latter-day reduction in lifespan? Would God interfere if the mortality from the epidemic or the fall in birthrate were too great? Mephisto wondered which side would be best served by male sterility – the Kingdom of the Air or the Kingdom of God?

NEW OUTBREAK OF ASIAN VIRUS REPORTED
Beijing, China by Anthony Wang, AP

October 26

"Officials in major Chinese cities have reported a record number of cases of a new highly transmissible virus. A spokesperson for the World Health Organization has designated the SARS virus as Gamma variant H3N7. WHO has issued an international travel advisory. Laboratories at Johnson and Johnson, Pfizer, and Moderno are racing to formulate yet another vaccine, which would be effective against the Gamma variant."

CHAPTER 11 – **Power Struggle over Succession**

London, UK

October 27   12:00PM

Council Member Icons

Figure 14. electron microscope image of Covid and HPV

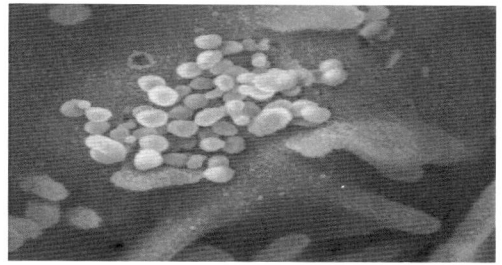

the Terrier   the Oil Derrick   the Crossed Sabers   the Top Hat

the Automobile   the Cargo Ship   the Pyramid   the Hammer

the Rugby Player   the Candlestick   the Hourglass   the Falcon

'*Q*' convened the third electronic conference of the Council of Twelve. All the member icons were illuminated, indicating that the members were entirely in attendance. A cash flow chart presented an account balance of over $252 Million. Next on the order of business was discussion of the progress of the global sterilization program. For the benefit of Council members, '*Q*' displayed news reports that describe the appearance of a new SARS virus variant. These reports originate from China's largest cities. There are also reports which come from Manila, Seoul, Taipei, Singapore, Ho Chi Minh City and Jakarta. A report from Rangoon states that the military had mysteriously closed Myanmar to international travel. The World Health Organization identified the strain of the virus and described the outbreak as occurring in epidemic numbers.

'*Q*' proposed, "The terms of our agreement with the *Terrier* are that release of funds held in escrow - $120 Million – would occur when the World Health Organization recognized the outbreak of SARS-CoVID-2. There is little doubt of the origin of this contagion. Therefore, the conditions have been satisfied. *Terrier*, your funds will be transferred by 1200 GMT October 28th, tomorrow night. Is there any further discussion on this matter?"

Most of the icons flashed. The computer selected the *Rugby Player*. "Simply marvelous. Congratulations on a successful start to a well-planned operation."

One by one, other Council members made similar comments, paying their respects to the *Terrier*.

The *Falcon* asked, "Are you on schedule to complete this 'business' and close the last set of clinics by March 30th of next year?"

"Yes, barring unforeseen complications," replied the *Terrier*. "The spread of this virus is meeting our expectations. However, the impact of the carrier-borne HPV will not be evident for months. You must also consider that the health authorities in some of the countries may deny that the HPV exists, or that it poses a problem."

The *Crossed Sabers* spoke. "Will you provide a report on the progress of these viruses at each council meeting?"

'No," the *Terrier* responded. "I need to limit my exposure for security reasons. Soon enough, the news services will provide a banquet of information. I will, however, provide anti-viral serum for the inoculation of those whom you select. You may coordinate these deliveries through the office of the Council Director. May I suggest that *time is of essence*, and that sooner is better than later. Also, I must limit the supply to reasonable quantities. I will ship the serum in accordance with the protocol that the Director has established."

That statement ended the questions for the *Terrier* for the moment.

The *Hourglass* inquired, "Presuming that the Council is still discussing the general subject of expenses, why are there no expenses listed for the Director's fees or his salary? Can you explain that omission?"

'*Q*' replied, "I have not omitted any Council expenses. I volunteer my time for the duties required. To an extent, you will have to trust that I have not skimmed from the general expenses, or that I collude with contractors. Allow me to pose a question to the *Terrier.*"

"*Terrier* – Do you intend to kick back any of that $120 Million to me?"

The *Terrier* was silent for a long moment. "No," he eventually responded. The *Terrier* was certain that he was being shrewdly manipulated, but he was unable to avoid the trap.

"Therefore," said '*Q*,' "I will take this opportunity for a vote of confidence which will be conducted by roll call. Indicate by responding 'aye,' 'nay,' or 'abstain'."

'*Q*' conducted the roll call quickly. One by one, the Council members all voted 'aye.'

The icon of the *Automobile* flashed. "Do you personally take adequate security precautions?"

"Yes, and increasingly so," '*Q*' replied. "I cannot elaborate."

The *Oil Derrick* asked, "Are there provisions for succession in the event of your untimely demise?"

"If you consult the Council Organizational Charter, you will find that the matter of succession to the Directorship is a by-law that is not subject to change by a vote of the Members. When or if I fail a vote of confidence, I will resign, and the Council will dissolve."

'*Q*' continued, "I have groomed a successor who has the requisite abilities for assuming the directorship. This candidate is not aware of the existence of the Council, or of any of its business. The candidate has assisted in preparation of video reports without knowing of their exact purpose. This individual knows that I am not given to idle pursuits. Each day I complete a required security precaution which includes this person. An encrypted message will be sent to my successor in the event of my 'demise,' whether it be incapacity or death. My successor will assume the director's position."

The next remark came from the *Pyramid.* "The Council must select the successor to the Director."

Other icons flashed, but '*Q*' muted them all. '*Q*' asserted firmly, "Each Member signed the Charter, which included the by-laws. At any time, a member may resign and forfeit the privileges of membership." Then '*Q*' reopened the communication channel for the Members.

The *Terrier* persisted. "I move that the Council adopt new by-laws concerning the Directorship and succession."

"The *Terrier* is out of order, declared '*Q.*' "The *Terrier* has been muted for the rest of this conference."

"Our next business," directed 'Q,' "is the Failsafe alternative to extinction, which is the Council's plan to build an orbital colony in space. My sources have informed me that the International Space Agency (ISA) is interested in privatizing their space station. Possession of the Space Station, and an international agreement to perpetuate new orbital coordinates at Lagrange point four, would advance the timetable for Failsafe. NASA, SpaceX, and ESA are on board with the plan to build the Lunar Gateway in sequential increments. Our intentions may also be temporarily veiled."

A thought suddenly occurred to Rothschild as to how to identify the 'Terrier's ardent supporters among the Membership. He knew from his private discussions with Sir John Clever (a/k/a the Terrier) that Sir John believed that the space outpost was critical to the survival of mankind. Surely Sir John had shared his enthusiasm for Lunar Gateway with his cadre of supporters among the Council membership. There is a way, Rothschild thought, to determine which members those supporters are right now, and gauge the level of their support for the Terrier. 'Q' had planned to conduct a voice vote after showing a video that had been specially prepared to promote the acquisition of the International Space Station. He had anticipated unanimous support for this proposal. But to flush out the Terrier's supporters, the time to take the vote was right now! Before the Terrier could contact his associates to influence the vote.

'Q' pursued the topic. "If a suitable buyer would agree to lease the existing laboratory facility and a habitat for a few scientists, the International Space Agency is likely to give the obsolete space station away, with a favorable location and its protected international status. The

Council, by proxy, would boost the ISS to the L-4 position, and construction would expand the existing station in stages. We would also be in position to remove dangerous 'space garbage'- orbiting flotsam - which would endanger our new colony. Presuming such favorable conditions, is there a motion to contract with the International Space Agency to purchase their space station?"

For what seemed an interminably long pause, there was no reply. Finally, from the *Top Hat*, "I move the Council to acquire the ISS from the International Space Agency, with the most favorable conditions that can be obtained."

'Thank you, Jeff,' Rothschild thought. Jeff Bezos was CEO of the world's largest internet merchandizer, owner of Blue Origin, and a close personal friend.

'Q' responded. "Very good. Is there a Second?"

"Second." The *Rugby Player,* owner of SpaceX. is another member on whom that Rothschild could rely.

"The Council is open for discussion," *'Q'* declared.

Only two Members raised points of discussion, and both were in approval. Then *'Q'* closed the discussion and put the motion to acquire the space station to a roll call vote.

"Members will vote 'aye,' 'nay' or 'abstain.' The *Terrier'*s vote will be recorded as 'abstain.'

"The *Oil Derrick* votes . . . "Aye"

"The *Crossed Sabers* votes . . . 'Aye"

"The *Top Hat* votes . . . "Aye"

"The *Automobile* votes . . . "Aye"

"The *Cargo Ship* votes . . . "Abstain"

There is one, Rothschild thought.

"The *Pyramid'* votes . . . "Abstain"

There is another.

"The *Hammer* votes . . . "Abstain"

A third "Terrier' supporter.

"The *Rugby Player* votes . . . "Aye"

"The *Candlestick* votes . . . "Aye"

"The *Hourglass* votes . . . "Abstain"

There is a fourth supporter.

"The *Falcon* votes . . . "Aye"

'There are four Council members definitely in the Terrier's bloc, and another. Those who abstained could not be certain of Sir John's position on the acquisition of the ISS; consequently, each took a neutral position. There may be one or more Members who are inclined to vote with the Terrier, but are more independent,' Rothschild thought.

'Q' summarized. "The ayes have prevailed, with none opposed and five abstaining. The Council will now view a video of the International Space Station. This video has been produced by our research team and is ten minutes in duration."

The Council members watched a presentation which showed video of an orbiting laboratory which had sustained man's presence in space for nineteen years. The space station represented the best efforts of human technology as of 2005, which was the year that the ISS was placed in orbit. The space station was hopelessly obsolete now and miniscule in comparison to the planned *Lunar Gateway* space station. Yet the docking station and the habitat provided a jump start for the first construction teams for *Lunar Gateway*. Even the mass of the old ISS had value. In one segment of the video, an astrophysicist described the importance and value of moving the orbital location of the space station to Lagrange point four. In another segment, a renowned attorney in the field of international law reviewed the applicable space treaties and the importance of the legal standing of the space station. The Council heard a scientist from the Jet Propulsion Laboratory on how 'space debris' in earth orbit had accumulated over time. The Council watched a simulation of space debris striking the ISS at orbital velocities, which demonstrated the danger of space junk, and the need to collect or disintegrate it. The video closed with archival clips of launches from SpaceX and Orbital Science Corp for Artemis I and Artemis II. These companies would perform the lion's share of the first stages of *Gateway*'s construction.

At the conclusion of the ISS video, *'Q'* announced that the upcoming *'Lunar Gateway'* video cost $100,000 to produce, and that the Council had employed the best talent in the 'special effects' industry. The twenty-minute video had been created in one hundred days. The result of their work was absolutely spell-binding. The video depicted a virtual city in space which was enclosed in a sphere. The sphere consisted of panels of air-tight solar panels. Within the sphere, there would be a breathable atmosphere and a holographic horizon which provided 'day' and 'night.' Private schools, sports, dining, and entertainment were planned. Gravity would be maintained that would be 70% of earth's normal gravity. Special medical services and longevity treatments would be available for aging citizens of *Gateway*. The financiers of the city orbiting at Lagrange point four boasted of plans for attractive 'green' habitats and hydroponic farms. A segment showed certain internal industries which would sustain the essential functions of the city, and another part of the video stipulated important space industries. The principal industry would be mining regolith containing water and rare earth elements – lithium, iridium, tritium, and Helium-3.

Engineers are building robots to mine the moon for the regolith. The ice-and-stone regolith will be screened on location at the Lunar South Pole. The refined ice, propelled by rail gun from the surface of the moon, is captured by special cis lunar cargo ferries and stored at *Lunar Gateway* for sale as fuel. The ice is obtained for a small fraction of the cost that would be necessary to carry fuel from Earth. Without this ice, space missions to Mars and the outer planets would be improbably expensive and the space journeys would be needlessly lengthy.

The small colony of astronauts would possess defense capabilities. Indeed, *the Lunar Gateway space colony proposed to be independent from Earth, and accountable to no authority except itself. One day soon, Lunar Gateway is where those privileged elite could flee if survival on the Earth came into greater jeopardy.*

CHAPTER 12 - **Clever Plot**

United Kingdom

October 27

Sir John Clever was not a man with whom one trifles. The Chairman of the Board of Directors of the pharmaceutical giant Pfizer-Welcom LTD was called by Time magazine the 'most influential' person in the world. Often decisions made at the level of Sir John's management determined the policies of the World Health Organization and the Center for Disease Control, not vise-versa.

To characterize Sir John's demeanor now as irate would be an understatement. Sir John Clever was always in control of himself and everyone in his dominion. From numerous past experiences in the push-and-shove of business and politics, Sir John had learned that revenge is a dish often best served cold.

Sir John had already made plans of his own to organize a tight cadre of leaders of industry and commerce who could manipulate the government and the military. When David Rothschild recruited his participation in the new Council of Twelve, Sir John was persuaded to join, even though Sir John's organization was operational, and some preliminary projects underway. The sterilization plan was a year old and would have begun in August regardless of the Council's vote. Joining the Council of Twelve had already reaped benefits. By virtue of the decision of the Council, Sir John would recoup $120 Million! Sir John had many strategies to make obscene profits from industries, markets, and even currencies that would be affected by sterilization and a plummeting global population.

Four of the Council's members were receptive to Sir John's private overtures. These men all controlled mega corporations and shared Sir John's vision of a New World Order. This New World Order would rationally be, to Sir John's thinking, a benevolent corporate dictatorship tightly controlled in a police state. These close associates agreed to their own pact of secrecy with the aim of dominating the agenda of the Council of Twelve. With only a couple of votes from Members outside their bloc, Sir John's strategies would prevail on any Council business. As the *Terrier*, Sir John led his bloc of colleagues on their own agenda.

Sir John felt that he had been bush-whacked on the matter of succession of leadership and then disrespected. The rule of succession was an organizational by-law. Rothschild was certainly smart enough to perceive that selecting a successor for Director by majority vote could be tantamount to a death sentence. This would be particularly true in the event of serious dispute, which was likely. The issue of succession of leadership came into discussion at the Council meeting unexpectedly, and Sir John made a tactical error in pressing Rothschild. When Rothschild stood his ground on the matter, and muted the *Terrier*, a man of Sir John's ego felt as though he had been publicly spanked.

Sir John believed that a demonstration of power was necessary 'to put Rothschild in his place.' It was not Sir John's intent to kill Rothschild, but to impress upon him the power and reach of the *Terrier*'s displeasure.

The incoming call on Sir John's secure line was expected. When Sir John connected, his agent said simply, "The bodyguard was compromised by our cash deposits in his personal account. The 'incident' will occur at the Royal Opera House at Covent Garden."

CHAPTER 13 - **Implications**

Kensington, UK

November 3

Nigel Cooper despaired over how his life had taken such a miserable turn. The position of bodyguard, one of three assigned by Avent Security Agency for David Rothschild, was the best job that Nigel ever had. Certainly, better than night watchman at empty factories or 'bouncer' at some raunchy late-night club. From the outset, Nigel was optimistic for a career with the Agency, which had a prestigious reputation.

The downward spiral had begun innocently enough. Nigel's position entailed diligent observation, but there were often twenty or thirty minutes at a time during his watch in which there was little to do. A friend had suggested internet casino, which was a card game that could be played on his smartphone. The game was a mindless diversion which helped to pass the time. A little wager on each game added a good deal of spice, as Nigel soon discovered. To gamble on the internet, one established a debit account with cash, and one played hands of cards until the cash was depleted. One could gamble for a few minutes, or for hours, depending on one's skill. Nigel was good at casino, or at least he had a good run of luck initially. After his first month of casino gambling, Nigel took a couple of his friends for a night out on the town at Nigel's expense, courtesy of his new job and his casino winnings. But that was history. Lately the casino debit account had become a sizable draw on Nigel's personal checking account.

Then there was the mystery of the unexpected deposits in his bank account. He received three cash deposits of $500.00 each over a short time. Nigel figured that there was some bank error which the bank would soon sort out. Then he had the 'brilliant' idea of increasing his casino wagers, so that he could cover whatever the bank would take when the errors were discovered, and his account reconciled. That idea fizzled when his gambling losses mounted, but he was having difficulty in his efforts to quit. Playing casino without a wager was useless; the excitement - the agony/ecstasy thrill - just was not there. Jamie Burton was the other bodyguard assigned by the Agency for the evenings with David Rothschild. Jamie noticed the casino games and gave Nigel a friendly warning to stop the gambling. Nigel figured that if he could not get a grip on his problems soon, it was just a matter of time before he was canned and on the dole. He would forfeit his nice flat in Kensington and must move in with his mother. Disastrous.

Then the big deposit appeared. $10,000. Nigel was making a cash withdrawal at an ATM when the ATM displayed a balance of over $10,000. He gaped at the transaction receipt in his hand. Nigel finally concluded that he was being manipulated. He would hear from somebody soon. And he did.

A text message without a caller ID advised him of a meeting at a small park near Grosvenor Square which would take place the next morning. With great misgiving, Nigel kept that appointment. In retrospect, Nigel knew that he should have 'come clean' with the Captain at the Agency. He would have lost his position and been banned from security work, but what is that compared with what could happen?

At the meeting in the park, he sat beside a dour-faced man who wore a brown woolen jacket and a derby hat. Using his knuckles, the man pushed a newspaper that lay on the bench toward Nigel. The newspaper covered a thick manila envelope. Nigel hesitated to open the envelope, but the dour man impatiently gestured for him to look at its contents. There were about two dozen photos. Large portfolio-sized photos. The first dozen or so were pictures of Nigel. Nigel at the door of his flat in Kensington. Nigel shopping at a neighborhood deli. Nigel at a pub that he frequented. Nigel standing beside Jamie Burton next to Rothschild's limousine. Nigel pumping petrol in the limo.

As Nigel peered at these photos of him, Nigel's face flushed with shame and embarrassment. He had been aware of no surveillance. As a trained professional, Nigel should have spotted the tail. But the next group of photos caused anger, then fear. There were photos of his mother and his younger sister; these photos appeared to have been taken recently. Finally, there were photos of David Rothschild. Nigel placed the photos back in the envelope. He was stunned and sickened by the implications.

"Listen carefully because I am only going to tell you once. Sometime in the next two days, go for petrol in the limo. Give us two hours' advance notice. Send a short text to this number. You will be given a local address of an automobile repair shop. Pull up to the left garage with the limo. If you do exactly what we say, nobody will get hurt. And there is another $10,000 for you when you bring the limo. Now sit here while I walk out of sight."

Nigel was left holding yesterday's edition of the Daily News.

Kensington, London November 4

Since Nigel worked the night shift with Jamie at Rothschild's luxury penthouse, he normally slept during the morning hours after work. But now sleep was next to impossible. His mind raced, thinking of courses of action, and discarding them just as quickly. There seemed to be no solution to his dilemma. He knew that he was a fool if he took David Rothschild's limo to the garage for 'repairs. The longer Nigel waited to turn himself in to the Captain at the Agency and face the consequences, the worse the outcome would be for him.

What to do? Nigel read his horoscope religiously. Often there was advice 'from the stars' that addressed his personal circumstances. 'Can't hurt to take a peek,' Nigel thought. He had a smart phone application that made the search quick and easy.

DAILY HOROSCOPE by Anton Mesmer

November 4

GEMINI (Oct 23-Nov 21): *Do not jump to conclusions or hurry to make decisions before receiving critical information. Have confidence in a successful plan.*

Amazing! Nigel thought. He put off further anguish over his situation.

Kensington, U.K.

November 5   4:30PM

Nigel worked another evening shift as bodyguard and endured another sleepless morning. He had remained awake last night only with coffee and

93

two pep pills. The use of *uppers,* as they are commonly called, was a violation of his employment contract. He would not pass the next random urine test. Nigel began to wonder if he would survive until the next drug test. With great misgivings, he texted the number that the dour man had given him.

"Going for petrol at 7PM." Within minutes, Nigel was given a local address.

Fueling the limousine was Nigel's duty, and no suspicions were aroused when Nigel arrived for his shift an hour early. Nigel wanted to avoid the possibility that Jamie might volunteer to ride with him. After checking in with the building supervisor and Rothschild, Nigel took the limousine for petrol and then to the garage. Nigel felt a measure of self-revulsion. He was indeed a fool. In a few minutes Nigel turned off the street at the proffered address.

At first the shop was closed for the evening. But when Nigel brought the limousine to the left garage door, the door rolled up, and Nigel drove inside. The door immediately closed behind him.

The dour-faced man still wore the brown wool jacket. He and an accomplice directed Nigel to drive the limo onto a hydraulic lift. The garage was dimly lit. Nigel stepped out of the limo, and the vehicle was elevated to chest height. The accomplice, who wore a mechanic's overalls, quickly set to work. He attached a small shop light to the frame of the limousine. Next, he removed from a large red toolbox what was clearly a bomb.

"Hold up there," Nigel protested. The mechanic ignored him. Nigel then faced the dour man. "I didn't sign on for any of this!"

The dour man laughed sarcastically. "What did you sign on for? A cruise ship to Bermuda? Listen up; you may be smarter than you look. This operation will be flawless, and nobody will be hurt unless you screw up. You have the detonator-signaling device, which is your cellular phone. Here is the plan. Rothschild has reservations for the opera at the Royal Opera House in Covent Garten in two days. Because of heavy traffic and limited parking, many of the limousines remain in special reserved parking. Usually, the routine is that the chauffeurs and bodyguards mostly loiter in the café until the end of the performance. Is that true?"

"Yes," said Nigel sullenly. Certainly, these people knew a good deal of the details of the opera schedule and Rothschild's security routines, Nigel thought. Nigel and Jamie had accompanied Rothschild to the Royal Opera House once previously.

"Our people want to make a 'statement' to Rothschild, but – let me repeat – nobody has to get hurt. After the performance ends, but before you and the other guard return to the limo, you dial the cell phone number for the detonator. The bomb destroys the limo, which should be empty. Because of all the big wigs at the opera, there will be a public spectacle. The police will be everywhere, all over the place. Rothschild will understand the message from our boss. Get the picture? Now let me watch you put the detonator number in your cell phone."

The fleeting thought came to mind that David Rothschild was a valued client of the Agency, and that Nigel was sworn to protect him. There was also the risk to Jamie Burton, himself, and others. What if these guys blew up the limousine on the way to Covent Garten? Nigel's brain was under the heavy influence of amphetamines and caffeine. He was also in his third day without adequate sleep.

The dour man lifted a slim valise and opened it, exposing currency in large bills. "Here is the $10,000."

Nigel nodded and took the valise. Any further recriminations were stifled.

"I repeat your instructions. You detonate the bomb at the close of the performance, but before you, the other guard, or Rothschild enter into the limo. Understand?" The dour man poked Nigel firmly in the chest.

Nigel numbly agreed. The events rapidly proceeded like a bad dream which he was powerless to end or even to affect the outcome. Nigel was sweating profusely and suffering from hot flashes. The handle of the valise with the cash almost burned in his hand.

The dour man continued, "Just so you know, there is a backup plan if you do not call to detonate. I will call the detonator myself if you have not done it by the end of the performance! It is up to you as the man at the scene to blow up the limo when there will be no casualties." The dour man gave Nigel a withering glare.

"You need to keep cool and act normally between now and the end of the performance on Friday night. Pull this off smoothly, and you will receive $20,000 cash in a lockbox at the Glasgow airport. You will have a new identity and a reservation for a flight to Brussels."

Nigel began to stammer incredulously, but the dour man interrupted. "Do you think that you could lay low and sweat out this little caper? You will need to take a little vacation. You will be met at the airport in Brussels. If all goes well, we may have use for you in the future."

CHAPTER 14 - **Destiny**

Kensington, UK

November 5

The instructions that Nigel had received in the garage two days previously seemed so simple at the time. But the exact timing of the call to the detonator was critical if no one was to be hurt in the explosion. Nigel was still suffering from sleep deprivation, and he was staying awake with uppers and coffee. He had twice called the box office at the Royal Opera House for a precise time for the show's conclusion. "Between two hours and two and one half, usually," was their response. 'Maddening,' thought Nigel. Then there was the difficulty of making the cell phone call to the detonator. Nigel could not think of an acceptable way to avoid Jamie's scrutiny. If, for example, Nigel excused himself to use the men's lavatory and then dialed the detonator, the explosion would immediately follow. That would be very suspicious. Any fool would conclude immediately a linkage. Nigel might as well hold up a sign saying 'Yes, I did it!'

Nigel convinced himself that he should call in sick. Then he could watch from a distance, say, with binoculars, to fine-tune the timing of the explosion so that no one would be inadvertently injured. But Nigel did not own a pair of binoculars, and there was no time to rent a car. There were problems with lingering at the scene prior to a bombing. Among other security guards, Jamie would certainly spot Nigel's old Ford. And what if the police, who were always on hand for security at the Royal Opera House, managed to cordon off the exits before he could escape? He knew that a Rapid Response Team was on duty; in fact, it would be impossible for Nigel to observe the exiting crowd without being recognized by one of the constables.

Nigel decided that the best remaining strategy would be to phone the Agency and call in sick. Then he could call Jamie while Jamie waited in the café for the opera to conclude. He would invent some excuse for calling and make sure that Jamie and Rothschild were not in the limo when he dialed the detonator. This escapade was becoming dicey and increasingly complicated. Nigel decided that he would blow up the limousine a little early and risk the displeasure of his blackmailers.

"Avent Security," the clerk at the front desk said. "How may I assist you?"

"This is Nigel Cooper. I am assigned to client #205. I am ill and cannot work this evening."

"You will need to speak with Captain Wilshire. Hold please."

Curses! Nigel thought. *That is the last person I want to talk with right now.* Nigel almost disconnected. But that would not do. The captain would call him immediately. A few moments later, a gruff voice said, "Mr. Cooper. I understand that three hours before your assigned duties, you are reporting that you are too ill to work. Do you realize that it is too late for the Agency to send a replacement?"

Nigel cringed. Just what he expected. "Captain, if I could work, I would," Nigel said lamely.

"Very well, take the time off that you need to recover. I will assign another guard to handle your duties. Be advised that the physician's assistant must conduct a fitness assessment before you may resume duty. Good day." The conversation ended about as badly as Nigel feared. A fitness assessment included a mandatory urine analysis. Nigel's career with Avent Security was over.

He thought about what he would pack for the escape to Brussels. Everything else he would have to abandon. There was just enough time to clean out his bank account. Nigel would do what was necessary tonight and flee to the Glasgow airport.

Royal Opera House in Covent Garten

November 6   9PM

Jamie Burton was in the café in the Royal Opera House when a call came. The caller ID was simply 205, which was David Rothschild's client number at Avent Security.

"Agent Burton," Jamie answered.

"Jamie, this is David. It is intermission now. The production has been brilliant, but I am only staying through the next aria. That would give us a jump on the departing traffic. Have the limousine in line at the entrance by 10PM sharply. I will call the limousine phone when I walk out of the theatre."

"Yes sir." The call ended. Jamie noticed that the battery indicator on his cell phone was low. He powered his cell phone off. He would not need it for a while, and he could recharge the phone in the limo on the return trip to Rothschild's residence. Jamie relaxed and ordered a cappuccino mocha grande.

Kensington

November 6   9:45PM

Nigel Cooper was panic-stricken. Jamie was not answering his cellphone, and Nigel knew that it was not smart to keep calling. In the event of an inquiry, there would certainly be a record of his missed calls. To compound Nigel's difficulties, the box office was closed as well.

'Come on, you idiot,' Nigel berated himself. *'Do something!'*

For the fiftieth time, Nigel checked his digital watch. He decided to call the detonator at 9:55. That meant waiting ten more minutes. The act of making the decision did little to calm Nigel's nerves. Nigel suspected that he was close to a physical breakdown. He could finally get the monkey off his back and relieve the pressure when he drove to Glasgow. Leaving London as soon as possible was becoming a better and better idea. He figured that he would start a life with a new identity in Europe somewhere. He had $15,000 in cash in one of his suitcases, with another $20,000 that he would collect from the lockbox at the Glasgow airport. Things could be worse. Feeling a little better about himself, Nigel checked his smartphone for a quick peek at his horoscope.

DAILY HOROSCOPE by Anton Mesmer   November 6
GEMINI (Oct 21- Nov 20): *You need to marshal your resources for a new venture. Use caution to avoid obstacles to your progress.*

Royal Opera House in Covent Garten
November 6   9PM

David Rothschild was leaving the Opera House to meet his chauffeur /bodyguard. He walked briskly to the main exit at the front of the theatre where he heard a familiar loud voice.

Figure 15. Limousine driver waiting at Covent Garten

"David, my good man! A moment, please!" It was Lord Beckham-Smythe, a member of the British Parliament and a wealthy investor in one of David's banking endeavors.

David almost groaned, as a conversation with Lord Beckham-Smythe would never take only a moment. David, however, stopped and easily assumed a patient and gracious smile. The two men spoke of family briefly, with David apologizing and promising to call at a suitable time. A glance through the double glass doors of the foyer showed that his limousine was near the curb. Lord Beckham-Smythe, however, seemed determined to detain him.

Just then, Rothschild's limousine exploded. The concussive blast shattered the glass doors and the foyer with a deafening impact that knocked Rothschild and Lord Beckham-Smythe off their feet. They both lay on the floor stunned and covered with glass fragments before being helped to their feet. Rothschild had partially shielded Lord Beckham-Smythe from the blast, and he was saying something excitedly which David could not understand. David was woozy, and he became aware of his own blood; he was bleeding from his nose, and he had cuts to the face

101

and hands from the glass. David lurched toward the gaping hole where the entry foyer had been, with Lord Beckham-Smythe tugging persistently on David's jacket sleeve.

They stumbled outside. What David witnessed was pure carnage. Two limousines and a taxi were burning, the dense smoke obstructing his vision. Close by were the bodies of two parking attendants in grotesque conditions; one was missing both legs. A limousine lay on its side. David could not determine which one was his limousine. His head hurt badly, and he slumped against the building façade. Sirens announced the arrival of police and EMTs at the scene.

A paramedic treated David, but he refused an ambulance ride to the hospital. David still clung to an irrational hope that his bodyguard Jamie Burton had somehow escaped the conflagration. As the minutes passed, this hope turned out to be ill-founded.

Beckham-Smythe insisted on sharing his limousine with David and delivering him to his residence. "You are lucky to be alive," Lord Beckham-Smythe declared. "You should take a leave from work until you are fully recovered." Then the Member of Parliament called Scotland Yard. Lord Beckham-Smythe had some connections there. Soon he was talking with someone high up the chain of command.

Rothschild sealed himself in his condominium with a detective from the 'Yard.' One detective had been assigned to his case who would also serve as a guard until David could make a revised security assessment. The detective who was waiting for him had already met with the building superintendent. After several questions, David's pounding head could take no more. He promised more cooperation later but now his body was shutting down. He showed the detective to a study where the detective could make calls and an adjacent bedroom where the detective could rest.

From his bedroom, David heard the detective talking with the building superintendent. The detective received and made calls in rapid succession. A thorough investigation was in progress. He heard the name of Agent Nigel Cooper mentioned repeatedly.

Rothschild lay on his bed. His mind demanded the release of his consciousness. As he drifted off, David pondered, "Was the bombing a random act of terrorism, selected based on the wealthy patrons of opera? Had some other scion of British society, industry or government been the intended victim? Or had he been personally targeted? Rothschild considered his purpose and the mission of the Council of Twelve with greater determination. The Doomsday clock was advancing closer to midnight.

England – Scotland Border, UK

November 7 3AM

Nigel Cooper raced north on the motorway from London to Glasgow. There was a thick fog and Nigel was driving faster than was prudent for the conditions. Amber fog lights illuminated the motorway, and his old Ford weaved its way around much slower traffic.

Nigel was listening to BBC reports of the bombing at the Royal Opera House in Covent Garden.

Nigel's cell phone chirped; he had a text message.

Who would be texting at this hour? he wondered. He slowed his speed slightly to read the text. There was no caller ID. The message was brief. "Mission accomplished. Ditch your cell phone."

Of course! His cell phone was a GPS for Scotland Yard! If the inspectors were onto him already, they were following his flight from London right now! The window rolled down, bringing cold damp air into the Ford. Nigel tossed the cell phone out the window with his right hand. He felt an irrational loneliness and naked without his cell phone. Nigel comforted himself with thoughts of the devices that he could purchase soon.

Nigel had been driving for hours. He had been mostly sleepless for going on six days. He was struggling to stay awake. He was nodding off and weaving between lanes. Twice the horns of motorists blared at him. *Let me see if a couple of cans of Jolt will turn the trick, Nigel thought.* He pulled off the highway at a petrol station with evening services. As he approached the door to enter the store, he saw a television monitor on the wall above the counter. The TV was displaying the photo on Nigel's driver's license. Nigel stopped in his tracks and quickly returned to his car.

Nigel had a bad case of the shakes, and he desperately needed sleep. Nonetheless, he returned to the motorway. Ten miles north of the border with Scotland, Nigel slammed into a bridge abutment at 90 KPH. As the horoscope forewarned, the concrete supports of the overpass became a fatal obstacle to Nigel's progress. Currency from a broken suitcase from the wreckage of the Ford blew in the wind.

CHAPTER 15 – **Licking Wounds**

London, UK

November 9

David Rothschild lay on a couch at home recovering from his injuries. The status of his personal and corporate security, however, required urgent re-evaluation. When Captain Wilshire with Avent Security offered to meet at Rothschild's luxury high-rise, David reluctantly agreed.

Figure 16. **Scotland Yard** *investigates (Courtesy of Four Square.com)*

Captain Charles Wilshire was already seated in David's spacious den. He rose when David entered the room. Wilshire was at least half a head taller, but the hand that gripped David's hand seemed to be disproportionately larger. Although Wilshire was in his mid-fifty's, he had aged gracefully, as befit an ex-officer in Her Majesty's S.A.S.

"You may recall that six months ago, we were scheduled to review security and assess your need for a heightened level of diligence," Wilshire said. "Your office personnel cancelled that appointment and never rescheduled."

David recalled the circumstances. At that time, there were many important projects, including the initial business of the Council. Security at that time seemed adequate, and he had given the appointment with Wilshire a lower priority. The implications of this lapse were becoming apparent. David nodded to acknowledge the truthfulness of Wilshire's assertion.

Wilshire reached into his valise for a few files. "Now there is a blot on the Agency's record - a record that previously was stainless. I want you to know that when I could not reschedule the assessment, I asked the Board to cancel your account. My request was denied for 'political' reasons. Now six people are dead, including both men who were personally assigned to you, and no one is stepping up to accept responsibility."

David could barely meet Wilshire's steely gaze. He was receiving a good military dressing-down, and it was well-deserved.

Wilshire handed a couple of bank statements to David. The statements showed recent activity on Nigel Cooper's account. David nodded.

Next came nine" x 12" photos from the police investigation. There were a dozen photos; a few were truly gruesome. David tried to push the photos away. His face was ashen.

"Look at them, damn you! Someone is not playing nice little games with you, and many other people may be at risk," Wilshire shouted. "It is just a matter of minutes before Scotland Yard comes calling! You better have your story together. I certainly will not be interfering in *that* investigation."

David's stomach heaved, and he barely kept its contents in his throat and sinuses. He lurched to an adjacent lavatory where he retched vigorously. About five minutes passed before David composed himself. When he returned, David had resolved to show no more weakness.

"I renew the Agency's offer to resign," said Wilshire firmly.

"No." David's response was barely above a whisper. His headache had returned vigorously, but David was determined to gain control of these matters of security.

"The Agency has identified individuals high up the corporate ladder at Pfizer-Welcom Industries, even Sir John Clever himself, is involved in what we believe is an assassination attempt. Do you have any reason to believe that Sir John or anyone else at the pharmaceutical giant may wish to do you harm or to gain financially at your expense or at the expense of some business that you control?"

Again, David nodded.

"I need to obtain some idea of the amount of money that potentially is involved in this dispute," said Wilshire. "Would you say ten million?"

"More, much, much more. I transferred $120,000,000 just this past week to a corporate entity controlled by Sir John. The funds are for transmission and control of a new virus related to SARS-Covid."

"After a business transaction of that size, Sir John is *unhappy* with you?" Wilshire was incredulous.

"I am just beginning to realize the extent of Sir John's greed and his ego. However, I plan to transact more business with Sir John. Much more business. Business critically important to climate control, world health, and population control," David said.

"In that case, we must make it expensive for Sir John to misbehave. I had a preliminary discussion with the Director on the possibility of going to 'war' with Sir John. A part of his corporate security is a Belgium firm with an unsavory reputation. The Agency has tangled with their thugs on occasion. The Belgians certainly turned our agent, Nigel Cooper. In the wreckage of his car, Scotland Yard found a false ID. Investigators found a reservation in that name with KLM Airline." Wilshire closed his agency file on Cooper.

"The Director thinks that Sir John's personal and financial empire is much more exposed than yours, by a factor of ten to one. For example, do you have a $30 million corporate yacht named *Our Merry Way*?"

"No."

"Sir John does. That yacht will be sunk at its mooring sometime this week. In a couple of weeks, Indian hackers – they are best in the malware business – will shut down anything that belongs to Pfizer–Welcom that is connected to a computer. Our Agency will tirelessly explore every opportunity to eliminate Sir John's Belgian thugs." Wilshire made no mention of further cooperation with Scotland Yard.

"For the Agency's upgraded security services, the Director requires a deposit of $8 million payable in two days."

David resolved to make his expenses for security an expense for the Council. Let Sir John and his allies chew on that at the next Council meeting.

Wilshire was scratching his head. "Why would you want to do any business with someone who shows so little regard for human life? Aren't you worried with becoming contaminated with some of Sir John's mischief?"

David had concerns on that matter as well, but his response to Wilshire was restrained. "My association with Sir John involves the future of the planet. Global climate change, disease and international rivalries put that future in jeopardy. Sir John is a piece in a complex jigsaw puzzle. To risk a second metaphor, Sir John is *the devil that we know*."

Chapter 16 – Russian-Ukraine War Expands

Figure 17. *Russian MiG-31 jetfighter/bomber*

MOSCOW (AP) December 10 — The Russian military said Thursday that it has deployed to Belarus, the country's westernmost Baltic region warplanes armed with state-of-the-art hypersonic missiles. This act comes amid soaring tensions with the West over Moscow's action in Ukraine.

Russia's Defense Ministry said three MiG-31 fighters with Kinzhal hypersonic missiles arrived at the Chkalovsk air base in the Baltic Sea enclave of Kaliningrad as part of "additional measures of strategic deterrence."

The ministry said the warplanes will be put on round-the-clock alert.

A video released by the Russian Defense Ministry showed the fighters arriving at the base carrying the hypersonic missiles.

Finland's Defense Ministry said Thursday that two Russian MIG-31 fighter jets were suspected of having violated Finnish airspace in the Gulf of Finland off the southern town of Porvoo, west of Helsinki. The Nordic country's Border Guard started a preliminary investigation into the incident.

The deployment of Kinzhal missiles to Kaliningrad as Russia's campaign in Ukraine nears the sixth-month mark appeared intended to highlight the Russian military's capability to threaten NATO assets. The region borders NATO members Poland and Lithuania.

Moscow has strongly criticized the deliveries of Western weapons to Ukraine, accusing the U.S. and its allies of fueling the conflict.

The Russian military says the Kinzhal has a range of up to 2,000 kilometers (about 1,250 miles) and flies at 10 times the speed of sound, making it hard to intercept. Russia has used the weapon to strike several targets in Ukraine.

Kaliningrad's location has put it in the forefront of Moscow's efforts to counter what it described as NATO's hostile policies. The Kremlin has methodically bolstered its military forces there, arming them with state-of-the-art weapons, including precision-guided Iskander missiles and an array of air defense systems.

"The events in Ukraine demonstrated that a clash with the collective West is a real possibility," Russian Foreign Ministry spokesman Ivan Nechayev said Thursday while emphasizing that a "direct confrontation with the U.S. and NATO isn't in our interests."

Speaking at a briefing, Nechayev said: "Russia as a nuclear power will continue to act with maximum responsibility" and "the Russian military doctrine envisages a nuclear response only in retaliation to an aggression involving weapons of mass destruction or in a situation when the very existence of the state comes under threat."

"We proceed from the assumption that the U.S. and NATO are aware where their aggressive anti-Russian rhetoric with an emphasis on a possible use of nuclear weapons can lead to," Nechayev said.

Figure 18. (Associated Press) Russian MIG-31s with Kinzhal missiles deployed to Kaliningrad

CHAPTER 17 – **Inviting Nuclear Brinkmanship**

The Pentagon's latest package includes up to $775 million of weapons and supplies from its stockpile.

Figure 19. The Defense Department is sending more rockets for the powerful HIMARS launchers the United States has provided to Ukraine, which have enabled strikes deep behind enemy lines. Credit...Corey

WASHINGTON — August 12. The United States is sending a new influx of arms and equipment that Ukraine will need for its counteroffensive against Russian troops in the country's south, the Pentagon said on Friday.

The Defense Department will also continue to send a steady stream of rockets for the **HIMARS launchers** that have been credited with destroying Russian command posts and ammunition depots, and other artillery designed to disrupt supply lines.

Taken together, the new shipment of up to $775 million of weapons and supplies from the Pentagon's stockpiles illustrates an emerging dual strategy: fueling Ukraine's immediate artillery fight, while also helping to build up an arsenal to support a counterattack near Kherson, in the country's south, which has yet to fully materialize.

The latest shipment includes 40 armored vehicles equipped with giant rollers to clear minefields ahead of any Ukraine ground operation, as well as 50 armored troop-carrying Humvees, 1,500 TOW guided missiles and 1,000 Javelin anti-tank missiles.

"The mine-clearing is a really good example of how the Ukrainians will need this sort of capability to be able to push their forces forward and retake territory," a senior Defense Department official told reporters on a conference call on Friday.

The Pentagon is also sending more high-speed anti-radiation missiles, or HARMs — air-to-ground weapons designed to seek and destroy Russian air defense radar. Military technicians have figured out how to integrate the American missile on Ukraine's Soviet-designed MiG fighter jets to help defeat one of the biggest threats to the Ukrainian air force.

The package also includes the HIMARS rockets, 16 105-millimeter howitzers and 36,000 rounds of ammunition, as well as 15 ScanEagle drones to help spot Russian targets and relay location information to the gunners.

For now, the United States has limited to 16 the number of HIMARS launchers sent to Ukraine, fearing that providing more would lead to burning through the Pentagon's stockpile of satellite-guided rockets and eventually endanger U.S. combat readiness.

Pentagon officials have emphasized in recent days that its resupply of ammunition for various artillery systems has now reached a regular, sustainable level that Ukrainian commanders can count on as they plan operations.

The shipment, the Biden administration's 19th overall to Ukraine, comes as fighting in Kherson, in the south, and the Donbas region, in the east, has ground to a standstill. A Russian offensive to seize Donetsk Province, part of the Donbas, has stalled — partly, American officials said, because Moscow rushed several thousand troops to the south to counter the anticipated Ukrainian offensive there and partly because of the effects of the HIMARS strikes.

"Right now, I would say that you are seeing a complete and total lack of progress by the Russians on the battlefield," the senior Pentagon official said, speaking on the condition of anonymity to discuss operational matters. "You're seeing this hollowing out of the Russian forces in Ukraine."

But when pressed by reporters, the official said that the Ukrainians lacked sufficient troops and combat power to drive the Russians from their defensive positions.

"We haven't seen a significant retake of territory, but we do see a significant weakening of Russian positions in a variety of locations," the official said.

The official repeatedly declined to comment on a series of attacks and other explosions in Crimea over the past two weeks. Ukrainian officials privately attribute the attacks to an elite Ukrainian special forces' unit operating behind enemy lines with the help of local partisan fighters. The strikes have shocked Russian commanders in Crimea, who thought their forces and weapons depots were out of reach of Ukrainian attacks, officials said.

John Ismay, *New York Times*, contributed reporting.

CHAPTER 18 – War in the Black Sea

Ukraine resists, damages Russia's Black Sea Fleet

Figure 20. The Dmitry Rogachev, a Russian corvette, traveling through the Bosporus to the Black Sea in February. Credit...B.Kara/Getty Images

September 24. Russia has replaced the commander of its Black Sea Fleet, the country's state news agency reported on Friday, following a series of setbacks that include a recent powerful strike on one of its Crimean bases and the losses of its flagship vessel in April and control of a tiny island in June that served an outsize role in Russia's naval operations.

The shake-up suggested the gravity of the setbacks to the Black Sea Fleet's operations. While there have been unconfirmed reports of similar major changes in the leadership of other forces, they have not been made public by the Russian government.

In a report by the state news agency, Tass, https://tass.ru/armiya-i-opk/1551/2727, on Friday, the new commander, Vice Admiral Viktor N. Sokolov, was quoted as saying that he had been appointed by the country's defense minister last week.

The comment came as he spoke to junior officers in Sevastopol, Crimea's largest port city and the base for the fleet since Russia illegally seized the peninsula from Ukraine in 2014.

"The Black Sea Fleet is participating in the special military operation, and is successfully completing all the tasks set for it," Admiral Sokolov, 60, told the officers, according to Tass, using the Kremlin's terminology for the conflict.

Admiral Sokolov, who served previously as the leader of the St. Petersburg-based Kuznetsov Naval Academy, Russia's top officer training school, added that the fleet expected to receive 12 new vessels this year, along with aviation and land-based vehicles.

He replaces Vice Admiral Igor V. Osipov, who had commanded the fleet since 2019. In May, Britain's defense intelligence agency reported that Admiral Osipov had likely been suspended following the sinking of the fleet's flagship, the cruiser Moskva. Asked about the report at the time, a senior Pentagon official went further, saying the commander had been dismissed.

Pro-Kremlin military analysts have cited the Black Sea Fleet as the weakest link in Russia's military effort. Since the start of the war in February, it has suffered repeated and embarrassing setbacks. Ukraine said it used Neptune missiles to sink the Moskva in April, a strike Russia has never acknowledged. It was the biggest warship lost in combat in decades.

The Black Sea Fleet is integral to the Russian war effort and has been crucial in Moscow's efforts to exert control along Ukraine's coastline, devastating Ukraine's economy. The fleet has also launched sea-based long-distance missiles to strike targets deep inside Ukraine.

"Most of Russia's naval victories have been achieved here by the Black Sea Fleet," Admiral Sokolov told the officers, according to a video report of the event by the local television network.

The change in leadership came as Ukraine has increasingly used sabotage and sophisticated longer-range weapons to strike Russian-held territory. But Russia has shown a robust ability to absorb losses and retains superior military might.

And it remains unclear exactly what toll the fleet sustained in the attack on its Saki air base in Crimea earlier this month, which Ukraine suggested had been carried out by special operatives and local partisans. Satellite images analyzed by The New York Times showed at least eight destroyed jets.

In a briefing for reporters on Friday, a Western official said that the attack had "put more than half of Russia's Black Sea Fleet naval aviation combat jets out of use." He added that "the Russian system is busy seeking to allocate blame for the debacle."

But U.S. officials disputed the idea that such a proportion of the fleet's aviation assets had been disabled. Recent Ukrainian attacks in Crimea have been significant, they said, and the explosions and damage Ukraine has caused have been real, including the loss of some fighter jets. However, they cautioned that the damage was not decisive, and the recent attacks alone were not enough to cause a shift in the war.

In June, Russian troops withdrew from tiny Snake Island in the Black Sea after repeated assaults by Ukrainian forces, limiting its control over Ukraine's shipping lanes.

— Dan Bilefsky, Ivan Nechepurenko, Steven Erlanger, Julian E. Barnes, and Eric Schmitt (source: https://www.nyt.com/live/2022/08/19/world/ukraine-russia-news)

CHAPTER 19 - **Vulnerabilities**

London, UK

December 28

David Rothschild had no doubt that global climate change was accelerating. He hired a renowned architectural firm to determine the extent to which the building in which he resided needed flood mitigation. David planned to remodel the building for security as well.

England – which was all that remained after BREXIT – was a sad shadow of the economic powerhouse that had been the United Kingdom in 2016. After the referendum to exit the European Union, Scotland withdrew from the U.K., and joined the E.U. Northern Ireland and Wales soon followed suit. England's economy tanked, with a trade war erupting with old partners in the European Union. Unemployment increased. The currency, the new pound sterling was devalued and remained at a thirty–year low against the dollar. Property values plummeted. England was in recession.

Rothschild purchased the building with his luxury high-rise for less than what it cost to construct the building in 2015. Some immediate improvements were undertaken for purposes of security. The primary purpose for David Rothschild was to display new realities for urban survival during global climate change.

Flood control specialists predicted that the river Thames would soon reach consistent levels of fifteen feet above flood stage. Buildings had to withstand near–hurricane winds. David sought to design a building that would serve as an example of what could survive and function during the gathering storms. David envisioned that much of London would become half-submerged like Venice unless radical solutions were adopted. Most of the mitigations for homes, such as shoring up and raising foundations, were expensive and well beyond the means of the average homeowner or landlord. Many Londoners would be rendered homeless.

David planned to lean heavily upon modern technologies. 'New Horizons,' as David renamed his latest acquisition, would not only be the tallest building in London but also be deemed the safest. He planned to implement a next-generation vertical isolation system that separates the building from the ground using more than one thousand huge shock absorbers. The system would work cooperatively with lateral isolation dampers consisting of huge sliding bearings. The massive-scale version of a modern car suspension would be effective against hurricane-force winds or earthquakes.

A partnership venture with Elon Musk, an innovative corporate billionaire, was promising. Wind turbines, generating power from winds in the North Sea, were transmitting power to satellites which could be relayed to New Horizons and other subscribers. Musk created a currency composed of kilowatt energy credits. Rothschild bought a significant minority stake in venture which was incorporated as KiloPower.

Clean water, water pressure and sewage treatment became important considerations for New Horizons. There were three firms who provided *green* proposals.

Access and egress from New Horizons were possible by boat, helicopter, or gyroplane. Musk partnered with Germany's largest gyroplane manufacturer to produce a battery-operated four-passenger short-take-off or landing (STOL) aircraft that could arrive and depart from the top of the building. David offered to finance the $250,000 gyroplane for all qualified tenants. He anticipated a design capacity for New Horizons of 2,000 residents. David expected to acquire adjacent properties at depressed prices.

David's brokerage was flush with cash from profits made in currency exchange, commodities and trading common stock. He correctly predicted the desultory effect that exit from the European Union would have on British commerce. He vowed to use those profits prudently; he founded a credit union charity to rescue Ukrainian refugees, extended family, friends, and employees suffering in the New Economy.

The hallway monitor showed Captain Wilshire and another man just as tall though not as large as Wilshire. After a retinal scan by a camera on the door, Wilshire entered the condominium and made his way to the den with the other man going to the terrace. Two other security contractors with binoculars were on the terrace making line-of-sight observations. The terrace, with its garden and spectacular view of the city skyline, had been off-limits to David since November 9. That was the date of the security reassessment. Because of the possibility of snipers, the terrace was enclosed. Special one-way glass panels permitted one-way viewing. The panels could also be used as monitors. Although most of their business was conducted by phone or video conference, the new security protocol required an electronic 'sweep' of a location where David met visitors or carried on conversations. Wilshire insisted upon conducting this sweep personally once each week after which Wilshire briefed David on security details. David remained silent during the sweeps.

With his task complete, Wilshire turned suddenly and waved a large finger in David's direction. "Don't assume that because you haven't heard the tiger roar, that the big cat is no longer hungry," Wilshire said sternly. "Sir John may wait until your next Council meeting when he will certainly attempt to seize control. Then you will be back to square one."

Wilshire paused, then said with bridled anger, "Have you forgotten the bombing at the Royal Opera House already?"

The pointed barbs irritated David, but he understood Wilshire's intent. There was need to strike hard and to strike soon at Sir John or the *'Terrier'* would become insufferable. Rothschild would not only forfeit the directorship, but he would also lose his life. Besides, David was tired of being restricted to the condominium.

"I have the resolve to see these important matters to an acceptable conclusion," David replied evenly.

The visitor walked briskly into David's spacious den. His stride had the lithe bounce and cockiness of a star athlete. Before David had grasped his hand or met his steely gaze, he surmised that this man was ex-military, retired Special Armed Services.

"Good afternoon. I am Morgan Trussler with G45. I will direct a tightly managed campaign to punish Sir John Clever for the bombing of the Royal Opera House in Covent Garden." In his speech, Trussler had the lilt of an aristocrat.

"I have provided certain evidence to the Chief Inspector at Scotland Yard. That evidence, along with the Yard's own findings, has convinced the Chief that Sir John is implicated in the bombing. But there is not enough evidence to indict, let alone convict a man of Sir John's stature. Also, if there was an active investigation by Scotland Yard, would that affect the work of the Council of Twelve?"

'Good question,' David mused. "A criminal investigation would jeopardize current business and cancel essential projects now in consideration. The police must not have further involvement."

"Unless we make a hash of this, Scotland Yard is willing to allow our agency to 'spank' Sir John. But the Chief Inspector was adamant – the only killings sanctioned are the Belgian thugs that Sir John retains," said Trussler. "The Inspector approved a 'tactical' strike on Sir John' private yacht. Additionally, the Yard's Cyber Crime unit will ignore complaints when computer networks at Pfizer-Welcom crash. The most valuable secrets of the pharmaceutical giant will soon be available to the firm's competitors – for a fee, of course. India's best clandestine hackers are not available for a song."

Both men maintained their erect posture. They possessed an air of confidence that was almost unnerving. David took comfort in the knowledge that these men were in his employ and not in Sir John's.

"When do these plans go in motion?" David inquired.

Both men glanced at their wrist watches.

"Synchronized operations at 2PM G.M.T. – One hour twenty-one minutes," said Trussler.

CHAPTER 20 - **Repartee**

United Kingdom

December 28   2PM

The Union Jack fluttered easily in the courtyard of the headquarters of the world's largest pharmaceutical firm. It was 2PM, GMT, December 28, when Pfizer-Welcom lost its digital consciousness.

At that time, urgent calls began to flood the technology help center, which was served by internet technology troubleshooters. Many confused and concerned P-W employees reported malware notices, which warned users not to shut down their computers. The most common notice advised that their valuable files were encrypted. To decrypt their files, the hackers demanded a ransom payment of $10 Million in bitcoin, an internet currency.

On the tenth floor, an IT administrator experienced an unrequested reboot which caused some inconvenience. He was preparing a software update for 110,000 employees when the reboot occurred. The administrator recognized the strong possibility of a massive intrusion of the P-W network.

In the open-office layout of the IT department, he heard cursing. He saw other laptop monitors reboot. On his own monitor, a malware message appeared, followed by the image of a cartoon character.

The administrator's training took over. Every minute that passed could mean the corruption of hundreds even thousands of personal computers linked to the network. There was only one stop–gap solution – an emergency network shutdown. The administrator raced to shut down the network. He jumped over a desk in route to the network server. The server scanned his digital recognition ID and approved his retinal scan. As he threw the lever for emergency shutdown, he recognized that the digital phones connected to P-W's international network were now dysfunctional.

Colin Ingrams was the skipper of *Our Merry Way*, a ninety-six-foot yacht which was registered to Sir John Clever. When not cruising from port to port at the whim and behest of Sir John, Colin usually relieved the first mate of his duties and had the luxurious craft all to himself. Maintenance and supervision of repairs required Colin's attention on M-W-F, so he relished Tuesdays and Thursdays as days off work. Sir John frequently took a weekend cruise.

Colin's routine on his day off was to sleep through the customary hangover, then shower, shave and dress up in his white captain's uniform. Colin always began the day at The Briny Cod which was just outside the Parliament Yacht Club.

"What a marvelous day for a holiday!" The skies were almost cloudless, which was a rarity for London at the end of December. Colin made his way to the sunroom, where he could watch the traffic on the river Thames. On an afternoon like this, Colin could also see Big Ben, the Tower of London and Parliament itself.

Maggie McGhee, a waitress well known to Colin, brought his customary dark and bitter ale.

"None of your besotted chaps have presented themselves yet," Maggie offered. There was a touch of Ireland in her speech.

"Tut-tut" Colin remonstrated in jest. "My esteemed colleagues are valued patrons of this exclusive establishment.""

"Their mere presence is an inducement to other fine gentlemen of distinction."

"Bah! I do not know why the proprietor tolerates your lot."

"The proprietor enjoys our singing."

"Your singing? You call your drunken ramblings singing?"

"Maggie, this is my day off, and you are giving me a headache."

"Alright, *Captain*." Maggie said sarcastically. "If this weather holds for another couple days, Sir John will want to take a cruise."

"You are probably right about that, Maggie,'" Colin acknowledged. "I'll have the Shepard's Meat Pie." The Skipper rubbed his substantial tummy.

"Why don't you get the fresh garden salad?" asked Maggie mischievously.

"Because a freaking salad doesn't absorb alcohol," Colin moaned.

KA-WUMP! A loud noise came from the direction of the Yacht Club's marina where *Our Merry Way* and more than seventy other expensive large craft were moored. After a few seconds, the alarms on several ships set off. Was one of those alarms what he feared?

"Argh!" Colin launched himself off the chair and lumbered his big heavy frame toward the dock where the yacht was moored. Even from a distance, something did not appear to be right…the mast was leaning and the whole ship was listing badly to port! And sinking. Slowly at first then more quickly. With mouth agape, Colin watched the $30 million yacht settle to the riverbed beside its mooring. From the ship's wharf, five skippers stared at the spectacle; the very top of the ship's conning tower was all that remained to be seen.

"Appalling!" James Mirthless exclaimed. "I heard an explosion," said the skipper of *High Tides*. *High Tides* was a yacht moored opposite *Our Merry Way*. "The blast was muffled; it must have occurred below the water level. Was the engine running?"

"No," replied Colin. His face was ashen.

"You'll have a rough time of it, mate." Another captain of a nearby yacht chimed in joining the forlorn group. They watched bubbles of air rise from the mostly sunken ship which had settled eight or nine feet to the riverbed. "The adjusters from the insurance company – their first suspect is always the skipper."

Colin felt the weight of eight sets of eyes. What to do first? Call the salvage company? The police inspector? The insurance agents? No. The call he had to make first was the one that Colin dreaded the most – Sir John. Colin had been the object of Sir John's wrath on two prior occasions. Regardless of the findings of the adjusters or the police inspectors, Colin knew that his employment with Sir John was history. That outcome was inevitable if for no other reason than the disaster that befell *Our Merry Way* took place during Colin's watch.

No, Colin was simply going through the motions, doing what must be done and hoping that Sir John would not *ban* him – keep him from obtaining other work in the maritime industry. Otherwise, Colin may end up washing dishes at The Birney Cod under the scornful supervision of Maggie McGhee.

He reached for his cell phone with great reluctance. With any luck, the electronic switchboard at Pfizer-Welcom would transfer his call into Sir John's voicemail. He autodialed. To Colin's utter amazement, his cell phone lit up with an animated image of a cartoon figure. With the manic trademark laugh-track ringing in his ear, Colin read the time: 2:10PM.

CHAPTER 21 - **In the Streets**

London

December 28 1 PM

Jacques Vander Weigel, native of Belgium, had grave reservations about the current assignment. Jack, as he was known to the Belgian contingent of Sir John Clever's personal security, was a veteran of mercenary postings in ten countries on three continents. Some of those missions had been extremely hazardous.

Jaime was scouting the perimeter of the home compound in an industrial park in south central London. Worthington, as the city borough was called, was a misnomer. However, the compound was the most defensible of the accommodations available to the mercenaries. Upon Jaime's return, he, Jack, Georges, and Pierre would depart for another worthless reconnaissance of David Rothschild's corporate office building and the residential building wherein Rothschild sought refuge.

The lack of realistic mission objectives was most worrisome. Each of the mercenaries in his security detail possessed an old wallet-sized picture so that if Rothschild "were encountered on the street, he could be identified and detained." Jack and his men received this order from the head of Sir John's corporate security. The Belgians' reaction was at first incredulity, then hilarious laughter. Was this multi-billionaire Rothschild chap likely to come out of a building ringed with armed guards in order to fetch his own fish & chips? Absurd!

The bombing at the Royal Opera House was a classic buggering. The explosive charge was too large, and the bomb should have been detonated in route to Covent Garten. Jack and his compatriots fled to

131

Brussels after that disaster with little desire to return to England. Curiously, Scotland Yard seemed anxious to place the blame entirely upon Nigel Cooper, Rothschild's guard. The Belgians returned to London two weeks ago. Jack harbored serious misgivings about continuing the contract in London.

In the seven weeks since the bombing, there had been many changes, none of them useful for the Belgian mercenaries. For Rothschild's team, there was a new and formidable player in the game – G45. G45 was highly skilled and politically connected. In the past week, the Belgians had twice scuffled with G45 agents; the first incident occurred on the street, the other in a pub. Jack was convinced that a more violent confrontation was imminent.

On the other hand, Sir John had done little to improve his own security. In Jack's view, Sir John's vulnerability to sabotage or even to suffer random acts of terrorism was far greater than Rothschild's exposure. Jack had said as much to Sir John's head of security without apparent effect.

Jack had multiple skills in martial arts. He was especially proficient in sav at (kickboxing). At his height of six feet, Jack found that his striking range with his feet was twice as far as he could reach with his fists. Like other fighting skills, a contest between two experienced fighters usually was determined by defense – blocking the opponent's kicks or punches, then counter- punching or counter-kicking.

Sir John had done little to protect valuable assets, and the counter-strike by David Rothschild's security forces was inevitable. Jaime returned from his lookout and hopped into the armored Range Rover.

The four mercs left the compound. Unseen by the Belgians, trailing their vehicle at a height of eight hundred feet, was a small electric surveillance drone.

Two kilometers distant, a drone operator contacted Morgan Trussler. "We have some activity at the compound. One armored car, four men. I am forwarding the GPS coordinates with a mobile marker."

Trussler activated the application on his cell phone. Instantly, a GPS screen appeared showing the moving position of the Range Rover. Trussler adjusted the scale of the display. A moving dot flashed on the GPS map.

"Sanction codes – one, two, three and four – set up signs blocking pedestrian traffic. THIS IS NOT A DRILL! Lorrie drivers begin circling the *Zone One* in five minutes – that is 2:51 PM. Agents Murray, Penders and O'Hara, move into positions," ordered Trussler.

Within minutes, four large *lorries* – straight trucks with van bodies – moved into the flow of traffic. Two vans circled a city block clockwise, one several meters behind the other. The other two vans performed the same maneuver in the opposite direction.

At two adjacent intersections of *Zone One,* four security guards posing as uniformed municipal utility workers set small yellow plastic barriers. The men warned foot traffic away from a solitary section of Euclid Avenue. Recent storm and flood mitigation had rendered solid steel–reinforced masonry up to eight meters on that stretch of Euclid. Rothschild had storm-proofed his corporate operations building.

"The trap is ready," Trussler announced. "Will Sir John's men take the bait and join us for tea?"

Trussler reached the drone operator on autodial.

"Send the driverless decoy limo from the garage at New Horizons."

The next call that Trussler made was to Scotland Yard.

"Chief Inspector, there may be some noise on Euclid Avenue uptown within the next half hour."

"This performance better be spotless," warned Scotland Yard.

"There will be no risk to the public," Trussler affirmed.

"Just don't leave a mess for us to clean up," said the Chief Inspector.

"Yes, sir, we will tidy up afterward." The G45 agent terminated the call.

Weeks previously, Sir John's corporate security planted a video surveillance sensor across the street from New Horizons. The primary purpose for this sensor was to keep an eye out in case Rothschild ended his self–imposed isolation. But for Sir John, an objective equally important was to intimidate David Rothschild. Sir John wanted David to know that his hired thugs were just waiting for David to leave the shelter of New Horizons. The surveillance also was to make sure that the Belgians were making their rounds.

Amidst of another meaningless uptown circuit of Rothschild's corporate building on Euclid Avenue, Jack's cellular buzzed.

"Oui?"

"Security One." That was code for movement by David Rothschild. "The limousine left the building seconds ago."

"Heading?" Jack knew better than to ask. He dreaded the answer.

"Undetermined.' The connection ended.

To his associates, Jack ordered, "Check your weapons. Security One." The coded order given was to capture or kill David Rothschild.

The Belgians were less than a kilometer from Euclid Avenue when Jack saw the Rothschild limousine. The limo turned in the same direction as the Belgian's Range Rover. Several cars separated the limo from the Range Rover.

"Perfect," said Jack. He kept the distance between the vehicles.

"How will this go down?" Jaime sensed that Jack was forming a plan. "There is no indication that we have been spotted."

"We will use the heavy traffic on this thoroughfare as a barrier. When the limo turns left on Euclid, I will cut across traffic and force the limo into the wall of the building. Since the right side of the Rover will be flush, we will all exit our vehicle from the left.'

Jack had orders for each member of the strike team. "Jaime will place the wheel chocks on the limo – one on a front wheel, one chock on a rear wheel. Georges and Pierre will deploy smoke grenades and keep watch of each direction on Euclid. You will accomplish these tasks within twenty seconds of our capture of the limo."

The intersection with Euclid loomed ahead, less than a minute away. The men in the Rover leaned close to hear Jack's additional instructions.

"After Jaime immobilizes the limo, I will drive a few meters to separate the vehicles. I will then use an electronic key fob to enter the limousine. Should I be unable to enter the vehicle, or if I am killed attempting to enter, Jaime will ignite a thermal grenade under the engine."

"No worries," said Pierre. "The limousine's occupants will be sauteed entries if they don't surrender."

Jack smiled grimly. "Pull down your ski masks now."

The mercenary had pulled off this maneuver in Prague in August 2021.

Morgan Trussler and Charles Wilshire stood in the situation room in the corporate office building just three floors above the unfolding drama. They were watching video feed live from the surveillance drone.

"Vans one and two – do you see a cream color 2020 Range Rover approaching the intersection? That is our target. In less than a minute, the Rover will pin the black limousine to the wall. You will have twenty seconds to block Euclid Avenue at the south end."

"Vans three and four – idle your lorries at your current position but do not enter Euclid Avenue until I give my mark. I repeat - wait until my mark to enter Euclid. When you enter Euclid, drive side-by-side to completely block the avenue at the north. *Exactly* like in our drills. Do not tip our hand by entering Euclid early."

The traffic light changed, and the black limousine entered the intersection. Jack floored the gas pedal swerving into the street lanes empty of oncoming traffic. The Range Rover roared past four vehicles that had been directly ahead taking the left turn on two wheels. About twenty meters onto Euclid, Jack slammed the Rover into the limo. With metal grinding against metal, the Range Rover forced the limo hard against the wall. Immediately, Georges and Pierre raced to their posts setting off smoke grenades to obscure the planned kidnapping. In seconds, neither end of the street between the buildings could be seen.

Jaime placed the wheel chocks on the limo, effectively freezing the vehicle in place. Jack moved the Rover away a few paces, while the tires of the limo screamed and smoked without effect.

Jack held the electronic key fob in one hand and an automatic pistol in the other. Jaime was in position a few paces away ready to shoot into the limo.

Jack pressed the release on the electronic fob. The bullet-proof windows were so heavily tinted that Jack could not see inside the limousine, but he heard the door lock pop open. Jack dropped the fob and opened the door. The limo was empty!

"Look!" Jaime pointed north on Euclid Avenue. As the smoke dispersed, two big vans appeared. Positioned side-by-side, the lorries totally blocked the north exit. As they watched, two men, one from each vehicle jumped out. They were carrying rifles.

Pierre was closest to the big vans. He fired a burst from his machine pistol from an extreme range; return fire quickly cut down the Belgian mercenary. He rolled once and lay still, bloody with legs a splay.

Wordlessly, Jack and Jaime sprinted for the armored Rover. As they piled into the vehicle, there was a shout from Georges. Two large vans moved into position at the south end of Euclid. They were trapped!

Georges sprayed the cabs of the vans with his machine pistol to negligible effect. Sharp shooters lay in prone firing positions. As Georges turned to run, he was struck by three high-velocity rounds. He fell on the concrete pavement, his blood pooling widely. He twitched once.

Jack's options were limited. A man emerged from the cab of a van who quickly hoisted a rocket-propelled grenade launcher onto his shoulder. He took aim at the armored Rover and fired.

Jack mashed the gas pedal, yanking savagely at the wheel.

Meanwhile, Jaime found religion. "*Mon Dieu, libere moi culpes*." He crossed himself. 'Maybe two seconds of contrition would be adequate,' he hoped.

For Jack and Jaime, the flare of the RPG and the end of their world were almost simultaneous.

Figure 21. Limousine sustains direct hit from RPG (vimbly.com)

Uniformed municipal utility workers used acetylene torches to cut pieces of the Rover into manageable sizes for disposal. Body bags marked 'John Doe' were loaded into a van which took them to the city morgue for anonymous cremation. Absorbent sawdust was used to lift blood off the pavement. The workers removed the portable yellow barriers that had been erected on Euclid Avenue.

Morgan Trussler phoned the Chief Inspector at Scotland Yard. "Everything is tidy."

Then Trussler called David Rothschild. "May I present your freedom and some just retribution. Sir John's empire has been badly shaken."

Rothschild congratulated the security agent on his successes, then became contemplative.

"Tell me, Mr. Trussler, must I adopt the tactics of those with whom I contend? Am I changed in subtle ways when I struggle against the likes of Sir John?"

"Whether you will be transformed by your association with Sir John Clever will depend on the strength of your convictions. As to the former question regarding tactics, that is the strength of the security team that you have retained," Morgan Trussler said. "We shall provide tactics that will limit exposure for you, your corporate assets, and any family and key employees that you care to identify. These are tactics of survival. However, our security services can only be as good as the intelligence that you provide. Captain Wilshire and I no longer feel the need to lecture you. You realize the consequences of neglect; you know from first-hand experience what the failure to observe security protocol may cost."

"What should I expect next from Sir John Clever?"

"An effort of reconciliation," said G45's top field agent. "Not a box of chocolates, nor a bouquet of flowers. Just an invitation to resume whatever business that is agreed upon. Look for a mediator to call you. The mediator will certainly be someone with which both you and Sir John conduct business."

"When do you think I should expect such a call?"

"Very soon. The digital assault on his corporations has brought his business to a virtual halt. Sir John will want you to call off the dogs."

"Just a moment. I have an incoming call on this private cellular from Barclays Bank. Sir John is on their board of directors. Do you suppose this call may be from a mediator?"

"Take the call," said Trussler. "Do not appear too anxious to settle your differences with Sir John. Under no circumstances," he advised Rothschild, "will you agree to meet with him."

CHAPTER 22 - **Bleak Prospects**

The private corporate conference in London on January 1 was chaired by *the Candlestick* in The Council of Twelve. From his keynote address came these remarks:

"Corporate Agri-chemical policy has a prime directive; to wit, to increase our product market while implementing policies that *achieve radical reduction in global population*. Specifically, our goal in India is to eradicate poverty by eliminating the poverty-stricken. The predicted drought has brought the opportunity to cull excess population. The plan is to distribute GMO drought-resistant seed for wheat and barley to subsistence farmers, then impose regulations that restrict use and marketing. The drought ensures that these farmsteads will fail. The subsistence farmers will starve or flee to the big cities for the food subsidies that our corporate associates sponsor. In the coming months, large numbers of these people will succumb to the impending SARS virus or other maladies. Because the majority of the poor in India are lower caste, there will be little effective political response to this strategy."

Andhra Pradesh state, India
Jailaun village
January 2   8AM

Syama Diksitar handed six consignment forms to Patel, an affluent villager. "Sign these forms for fast-track title transfer," the government official said. "The second form lists the cultivation and marketing requirements of the Agri-chemical corporation."

"Were all these properties abandonments?" Patel flipped through the forms, looking at addresses.

"Yes. Bodies were found in one farmstead. A state cleanup crew was working there yesterday."

"I do not see the Gupta property. I presume he is still on his farm, urinating on the dust," Patel laughed.

"The Gupta farmstead is on my list to visit this morning," the government official replied humorlessly. "Since you own two adjacent properties, it is in the government's interest to offer the acreage to you first, should the property become available," Syama advised. The government official waited patiently to conclude the transaction.

Patel signed the forms. He reached into a money belt which was almost hidden by his girth. He handed Syama an envelope. "It's all there," the villager said.

"I'm sure," responded Syama.

Figure 22. Andhra Pradesh State, east coast of India (en.wikipedia.org)

Andhra Pradesh state, India
January 2   9AM

The sun beat heavily on the Indo-Gangetic Plain. This area, in good times, was a breadbasket of the world. But these were not good times. An early morning temperature of 28 degrees Celsius doubtlessly preceded another dry scorcher. A small motorcycle bearing a government license tag scattered dust from a trail that led out from the village of Jailaun.

Syama Diksitar, a government official from the Indian Dept. of Agriculture, Andhra Pradesh state, wiped at the dirt on his googles. As far as he could see, stunted sprouts of wheat testified to the long drought which persisted in central India. Half of India's population are engaged in agriculture, mostly subsistence farming. On this trip, Syama needed to report how subsistence farmers were faring. The welfare of the farmer that he intended to visit would provide a good indication. Often a glance or two around the homestead would be all that he would require to make his assessment. Nonetheless, the state required that its forms be fully completed. If anything was true, Syama was a consummate bureaucrat.

Syama was well equipped to interview his clients in their native languages. In addition to East and West Hindi, Syama spoke English, Telagu and six other tribal dialects. Hundreds of languages in India were spoken by 60,000 or more people. Usually, one language bore no relation to another and were mutually unintelligible.

Syama took special interest in Ranjun Gupta, a farmer whose plot of cultivated land was near the village of Jailaun. Jailuan, typical of nucleated villages, consisted of inner clusters of middle-caste residents

143

which were encircled by farmers of lower caste. Ranjun was lower caste. He spoke Telagu as did most of the residents of Andhra Pradesh state.

Syama was aware that most subsistent farmers did not rotate their crops. There existed a steep reduction in the market price between wheat and barley; thus, crop rotation was a luxury that Ranjun could not afford. Soil nutrients were depleted continually, crop after crop, year after year. Rain, when it did arrive, was in monsoon quantities which washed soil nutrients into irrigation ditches and away from short-rooted wheat sprouts. For subsistence farmers, excrement - mixed with that of oxen - is the only affordable fertilizer. In recent years, the state distributed some fertilizer and seed grain that was drought-and-pest resistant. But inconsistent rainfall, failing wells due to falling water tables, and pestilence lead to crop failure on a widespread scale.

Syama viewed Ranjun Gupta's plight as typical of Indian subsistence farmers. His farm was ancestral, subdivided over several generations. The farm consisted of his modest home and an area of cultivation that barely measured two acres. His family, five in number, lived in a simple, one-story mud structure with a flat thatched roof. The house is called a *kacha*. The kacha was shared with an ox which was essential for tilling the soil and carting water.

As with most Hindus, the Guptas were vegetarian, so the family ox was never considered as a potential source of food. Hindus revered oxen and often allowed them to roam freely to feed. Ranjun's ox was essential for the purposes of tilling the soil, spreading seed and fertilizer, carrying produce to the mill, carting water, and a dozen other mundane household tasks.

Upon arrival, Syama dismounted from his cycle and looked about. He was alarmed when he saw the emaciated frames of Ranjun's two young sons. Their skin was stretched tight over ribs, and they had matchstick arms and legs. Their stomachs were slightly bloated. They were playing listlessly on a hard-baked surface in front of Ranjun's kacha. It was apparent to the government official that Ranjun's family was starving. These boys were Ranjun's pride and joy. If this was the condition of the boys, how were the wife and daughter faring? Syama reached inside a pouch on the cycle and gave each boy a government-issued protein bar. The eyes of the boys sparkled, and they nibbled gamely at the treat.

Figure 23. Courtesy of Britannica.org

Still there was still no greeting from inside the kacha. There was no sign of Ranjun. But off in the distance, obscured by heat rising from the fields of dirt, he could see a half-naked dark-skinned man toting two large metal buckets. The buckets were suspended from a wooden shaft that the man carried atop his shoulders. As Syama watched, the figure staggered under the weight of his load. Syama ran to the man's aid. In a few minutes, Syama was carrying the heavy buckets of water. Long before they reached the kacha, Syama wondered how the thin man could lift the buckets, let alone carry them for the five kilometers from the Jaira well. This task was a job for the ox. Where was the ox? Syama assumed the worst. *The ox must be dead.*

Coming to Ranjun's aid had curtailed the traditional Hindu greeting. When the men reached the kacha, Syama more thoroughly examined the physical condition of his client. Ranjun was dehydrated and exhausted. Was he delirious as well? Ranjun could barely be understood. Syama had visited Ranjun several times over the past few years; Ranjun knew well who Syama was. Was the man trying to call him *Vishnu*?

Ranjun was shaking; he fell to his knees in prostration. Ranjun's world was coming in and out of focus. The mantra he had begun at the Jaira well had served its purpose; Ranjun had toted close to his own weight without hydration or rest for five kilometers. This was a super-human accomplishment, especially considering how close the man was to starvation. Then, in his time of trouble, *avatara* – an avatar or incarnation of Vishnu – had appeared in the form of Syama, the government official, to set matters right.

Ranjun's wife Indira finally appeared, pulling out two mats and imploring for the men to sit. Shortly she delivered a pot of tea and two ceramic cups. Indira was wearing a traditional Indian garment that partially hid how terribly thin she was. Her hands, however, were cadaverous, and her eyes were sunken into her skull.

She disappeared back into a bedroom in the kacha.

Syama produced three more energy bars, which Ranjun gratefully accepted. One bar for Ranjun, one for Indira, and one for Sari, a daughter who of late stayed in the house. Under the circumstances, Syama did not inquire after the daughter. It was frequent practice in lower caste families that when rations were tight, a daughter received half portions.

Syama quickly got to business. "When did the ox die?"

In Ranjun's state of mind, the question was further proof of the avatar's clairvoyance. "Almost a month ago," Ranjun replied sadly.

This fact made several other questions superfluous. Since the river was too low to feed the irrigation canals, and rainfall was at historic lows, the death of the family ox was a serious blow to the sustainability of the Gupta farmstead. There were no rupees to buy another ox. Tilling the soil, planting the seed, and carting of water for the wheat had to be done by hand. That scenario was impossible to maintain.

"Have you come to tell me about any new aid programs from the government?" Ranjun asked hopefully.

"Other than a new hybrid seed and fertilizer sufficient for two acres, there are no aid programs that I can qualify you for. Next week the State truck will bring the new hybrid seed in exchange for the old seed. Show me the seed for next crop."

Ranjun was ashamed and could not look at the government official. "I gave the seed to Indira to cook with the barley soup." Both men knew that this disqualified the Gupta farmstead from the state seed exchange. Surely *Vishnu* would forgive.

Syama thought for a moment. The State Department of Agriculture encouraged migration from the subsistence farmsteads. The much larger surrounding farms would absorb the Ranjun's abandoned acreage. "You are aware of the food distribution in Nagpur and Hyderabad?"

"I have heard of such charities. I could walk away from my ancestral property, selling it for a pittance to Patel, like my brother Sonjay did last year." The upper-caste Patel family lived close to the core of the nucleated village.

"Where is Sonjay now?"

"Hyderabad." Hyderabad was some ninety kilometers southeast. Prospects there were uncertain. Some migrants from the farms could afford to move into slum flats, often sharing space with earlier migrants from their native villages. Others had no recourse but to find shelter among the clusters of dwellings called bastis. Bastis are makeshift dwellings which are commonly found along railroad yards, outside the walls of factories, along the banks of rivers or wherever else the urban authorities permit. Either option would result in a reduction in caste for Ranjun and his family, should they survive the trip.

"A man does what he has to do," was Syama's parting advice. Syama stood to leave, taking one last survey of the dacha common room which, he knew he may never see again. A wooden Shiva stared back at Syama from a niche in the wall that served as a family alter. The government official pressed his hands together and with a Hindu benediction wished Ranjun well. He left never to return.

Ranjun knew that the road to Hyderabad presented many difficulties. Possessions would have to carried or drawn by cart. He would not be able to resist the thieves that would be encountered. It was unlikely that Indira and the boys would survive the trek. Abandoning their bodies along the road to be scavenged by birds and rats would not be honorable. Prospects were bleak.

Late that afternoon, the daughter Sari died. A river ceremony was out of the question. Ranjun gathered such flammable materials as he could find. Sari's tiny body was placed on top of a pyre in front of the dacha and completely burned. Ranjun spread the ashes in the dry irrigation canals where he believed that someday Sari's essence would find its way to the eternal river.

Ranjun remembered the last words of advice of the government official. He instructed Indira to prepare the last of their barley for soup. After the sparse meal, Ranjun's concern were important *samskaras*, impending life-cycle sacraments for the remaining family. The sacraments are traditional Hindu rites intended to make his family fit for the next stage of life. Although his sons were young for the *upyanana* initiation, the ceremony was essential for his sons to enter the spiritual community of *dviga*, the twice-born. After offering incense and ritual prayers honoring Shiva, the Hindu god of destruction, Ranjun believed that he understood his role in the renewal of the universe. He was convinced that his respect for the Hindu deities guaranteed reincarnation at a higher caste.

That night, Ranjun waited until his family slept. Tearfully, Ranjun slit their throats. A handful of matches started an adequate fire. Then, summoning all his remaining strength and will, he fell on the sharp blade. Ranjun's last conscious thought was his confirmation of entrance in the celestial community of *dviga*.

CHAPTER 23 - **Northwest Passage**

Kingdom of the Air

January 2

Legionnaire emeritus Yuan Li was summoned for a conference with Lucifer. As usual, Yuan Li transported into Lucifer's presence without further notice. Yuan Li knew to come immediately; delay would mean delivery via the tractor beam which was always humiliating. In that Yuan Li was observing an important meeting of the Chinese Central Committee, the interruption was annoying. The Chinese government deliberated responses to the SARS viral pandemic and the United States' proclamation of *Pax Americana*, enforced by SCOLD. Lucifer's requirements, however, trumped whatever duties in which a demon might be employed.

There was a familiar background venue. A bald-headed Lucifer with a halo and white robes stood in front of a hypothetical heaven's gate. The gate was composed completely of jewels and was suspended in mid-air. Cirrus clouds moved slowly in the distance. Lucifer masqueraded as St. Peter. In perpetration of the doctrinal myth, as head of the Church, St. Peter ruled on the admissibility of souls into heaven.

Yuan Li was not surprised when Mephisto, another high-ranking legionnaire, materialized. Although the two demons had shared a mutual animosity for thousands of years, for the past year the strategies of Lucifer had required close coordination of their responsibilities. Another strategy session was obviously forthcoming.

"The viral pandemic is growing rapidly and spreading from the targeted cities. In the coming weeks, the effectiveness of the virus in sterilizing surviving victims will be evident. *The Council of Twelve's plan to reduce the earth's population to five hundred million proceeds admirably.* However, the deaths from SARS and attendant diseases, even with the drop in the global birthrates, are insufficient to meet target population reduction," Lucifer said.

"What about warfare?" Yuan Li asked. "What about WW III? No telling how many humans would be killed in another world war."

"That's precisely the reason why world war is ruled out," Lucifer responded. "Human technology has increased to the point that an all-out exchange of nuclear ICBMs would result in extinction of life on the planet. Mutual Assured Destruction. The Kingdom of the Air wants to preserve mankind in rebellion, not totally destroy it. Demons gloat about the destructive capabilities of Space Command's Orbital Laser Defense (SCOLD) and extoll its virtues as if it were a completed Tower of Babel. Although the first practical use was for planetary defense from an asteroid, the real purposes of SCOLD was to prevent all-out thermonuclear war. Mutual Assured Destruction is now a fear of the past because the space-based laser weapon system gives the United States too great an advantage."

This was old news to the two crafty legionnaire demons.

"SCOLD doesn't rule out conventional warfare," opined Yuan Li.

"That fact is especially true if the world powers fight *proxy* wars with limited objectives," added Mephisto. "Ukraine is a proxy for the

United States in the current conflict with Russia and their Donbas partisans. For another example, if a war erupts between Israel and Iran, I do not have to be Nostradamus to predict the U.S. would support Israel while the Russians would support Iran. There is more likelihood that such a conflict can be contained in the Middle East."

'That was strange,' Yuan Li thought. Normally, Mephisto only took potshots at Yuan Li's suggestions. Was an age-old conflict mitigating? Or was Mephisto just setting Yuan Li up for a colossal failure?

Lucifer regained control of the meeting. "I have other less capable legionnaires charged with instigating limited conventional warfare. I have broader, more ambitious plans for you both. You can seize grand opportunities that global climate change provides."

"Yuan Li. You have expanded Chinese influence, or should I say *neo-colonialism*, deeper into southeast Africa. The timing is superb. Storms with record flooding have destroyed traditional agriculture in wide parts of Mozambique, Zambia, Tanzania, and Zimbabwe. The flooding and relief expenses have bankrupted these countries. The changes in climate now taking place favor the cultivation of rice. What people know more about rice than the Chinese? Simply offer economic assistance loans to the southeast African nations of the same type that the Chinese recently made with Kenya. Once again, there will be no competition from Western banks. Make these loans short-term and secured by agricultural properties that are currently swamped. Then, offer loans for infrastructure which stipulate that Chinese construction companies build hydro-electric facilities, riverside levees where needed, irrigation canals and planned

153

communities for Chinese workers. Make loans for free food distribution in the big cities where the poor eke out a living, including those who migrate from the muddy farmlands. Most will fall prey to the viral epidemic or other misfortunes. When the governments default on the terms of repayment, simply foreclose."

This was advice expected from their demonic mentor. But what followed at first seemed out of character, until his Legionnaires listened carefully.

"I admonish you to be generous. Only foreclose on half the properties right away. Give the market a chance to rebound," Lucifer said. "Establish corporate models for successful big farms that are mostly automated. Foreclose on the rest six months later. At the outset of your business together, your new African clients will thank you for their rescue."

"Mephisto – I wouldn't authorize a magnificent new responsibility if I had any doubts about your ability to supervise current projects and succeed in a new joint venture with Yuan Li."

"Yuan Li – Does China chaff under the burden of U.S. tariffs? Is it possible to divert Chinese merchandise to other markets? Global climate change has opened up the long-sought Northwest Passage through the arctic regions of Canada. The Passage links merchant vessels sailing from ports on the East China Sea through portions of the Arctic Ocean clear to Europe. Conversely the Passage will open a direct trade route from Europe to markets in China.

"Global climate change offers opportunities for developing mineral resources in northern Canada and in Greenland that previously were inaccessible. These mineral resources in northern Canada include the Athabasca Tar Sands which may contain the world's largest reserves of oil and natural gas. Financing for Chinese equipment and mining technology will be welcome in developing the Athabasca. In Greenland, the mineral resources include uranium and rare earth elements. China can control the extraction and marketing of the resources thus ensuring Chinese domination of materials essential for PCs and other electronic devices.

"Global climate change also presents a time bomb for Canada in the Hudson Bay Lowlands and the Athabasca River Basin. The increased annual rainfall keeps the River at flood-stage; the alluvial plans in Alberta are a huge swamp. The world's largest deposits of peat moss lie within what is now mostly permafrost tundra, which is thawing at an alarming rate. The provincial governors of Ontario, Manitoba, Alberta, and Saskatchewan recognize that current conditions exist wherein the thawing peat moss are releasing vast quantities of methane and carbon dioxide. No longer can advanced methods of greenhouse farming trap these gases before they are released into the atmosphere. The presence of over 100,000 freshwater lakes, persistent winds for turbines, solar power and the temperate conditions predicted for these provinces could be ideal for GMO rice and harvesting edible crickets. Yet, the clock is ticking away for such solutions for the peat moss conundrum."

"The Chinese can supply an important piece to this puzzle. Can you provide an advantage for the Chinese in trade talks with the Canadians?

Figure 24. **Northwest Passage becomes reality as sea levels rise.**
(Britannica.com)

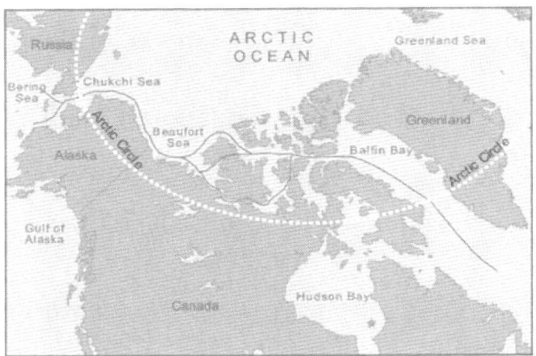

Canada had a very favorable experience with large numbers of Chinese emigres from Hong Kong in the first decade of the twenty-first century. But could the increased trade between Canada and China due to global climate change drive a wedge between long-time allies like the U.S. and Canada?

"Mephisto and Yuan Li, employ whatever devious tactics are necessary to bring China and Canada together in a Northwest Passage Partnership."

"Mephisto, provide compelling incentives for the Council of Twelve to organize the agricultural revolution and energy resource development in Canada. Promote investment proposals for a large desalination facility in Manitoba on the Hudson Bay. Due to the poisoning of the Athabasca River Basin by strip mining, fresh water for agriculture will soon be a luxury in Midwestern Canada.

"Time is of essence," Lucifer insisted.

"China must place Canada into its sphere of influence soon before climate change delivers waves of affluent immigrants from Europe," Lucifer warned.

"Why would affluent Europeans wish to move to Canada?" asked Mephisto.

"Once again, global climate change will create opportunities for those of means and disaster for those without," Lucifer affirmed. "It is estimated that one-fourth of all fresh water is frozen in the great ice sheets and glaciers of Greenland. When the melting of all the ice accelerates, the temperature of the water pouring into the north Atlantic will drop by 3-5 degrees. Already, the salinity in the waters surrounding northern Canada has been significantly altered. The delicate balance that supports a warm subsurface cross-current and a corresponding warm mass of air - a combination called the Atlantic Gulf Stream - is collapsing. The loss of the Atlantic Gulf Stream will impact weather in Europe tremendously. The consequence of a much colder Europe is economic disaster."

Mephisto recognized that Lucifer, as Lord of the Air, possessed six millennia of experience as a master of planetary weather conditions. The demons were eyewitnesses to both the creation of the earth and the catastrophic flood. The destructive fury unleashed at the onset of the flood was an all-to-vivid memory for Lucifer's trusted demon. When the very plates of earth's crust were violently dislocated with worldwide volcanic eruptions, Mephisto feared for his very existence. As Lucifer shared his insights on global climate change – which Mephisto had no reason to doubt – the demon reconsidered how certain delicate balances had occurred after the Flood. *Had God guided the formation of these delicate balances to protect a small and fragile mankind?* Mephisto dispelled the notion. Those thoughts led to madness.

"What an irony that Europe should prepare for the deep freeze," said Yuan Li, "when the average global temperate is slowly increasing! What do you suppose would be the financial consequences of a much colder Europe?"

"Agriculture and tourism will suffer acutely. Millions of jobs will be lost. The cost of energy will skyrocket. We are talking about hundreds of billions of dollars," Lucifer predicted. "Citizens with the means to leave Europe will do so. Immigration to Canada is very restrictive and favors those with money. For the affluent, immigration to Canada will become attractive."

"This plan, properly developed, will weaken Europe, and drive a wedge between two long-time allies – the United States and Canada."

Council of Twelve intranet communique between the *Candlestick* and the *Oil Derrick*

Re: Use Chinese capital for development of energy resources and Agri-chemical markets in Canada

"The prime directive is to achieve radical reduction in global population."

"The corporate Agri-chemical goal in Canada can build a dominant market share in the provinces that will emerge as the world's most productive farmland. Peat moss deposits can be converted to bio-mass, forestalling release of vast quantities of methane and carbon dioxide."

"Big Energy can develop fossil fuels, wind, and solar energy resources for new markets in Canada, China, and Europe. However, the Athabasca fossil fuel reserves, the world's largest, cannot be extracted without venting of greenhouse gases – methane and carbon dioxide. *The consequence of poor resource mining and mismanagement can very well push the fragile air and marine environments past a tipping point from which no recovery is possible.*"

"These goals rationally create a cooperation that is mutually inclusive. The Council of Twelve must lobby cooperatively for a Northwest Passage Partnership between Canada and China for purposes of facilitating trade and development of energy resources in the provinces of Alberta, Ontario, Saskatchewan, and Manitoba."

"The new trade pact will be perceived as a threat to the impending Alliance Empire in which the United States will dominate. The Empire will become isolated and will radicalize. The Prime Directive will face little opposition."

CHAPTER 24 – **Great Britain called to Task**

Courtesy of Somini Sengupta
Global Correspondent, Climate

Charles at the 2021 United Nations climate change conference, when he was Prince of Wales. Pool photo by Yves Herman. Figure 25.

The crown and climate

The new king of Britain, Charles III, has long been outspoken on conservation and climate change. Charles leads the institution he represents, and his country's efforts to tackle global warming.

The weight of history

The Industrial Revolution was born in England in the 18th century and so, in a sense, was climate change, as the burning of coal, oil and gas produced vast quantities of greenhouse gases, warming the Earth's atmosphere.

Britain led that transformation as an imperial power. Key to its dominance was its ability to extract natural resources from its network of colonies around the world.

The past matters. Many of the countries formerly colonized by European powers are today impoverished. They have few resources to deal with the hazards of climate change. The Intergovernmental Panel on Climate Change spelled that out in a report this year, citing colonialism as having exacerbated the vulnerability of formerly colonized people.

Colonial Britain's most prized possession, British India, has become the independent republics of India, Pakistan, and Bangladesh. Their people have been pummeled by extreme heat, erratic monsoons, melting glaciers and sea level rise — all telltale signs of climate change.

Important to remember: Before the Industrial Revolution, Britain prospered as one of the world's most prominent slave-trading countries.

What Charles has said

Charles has acknowledged the depredations of both. On a visit to Rwanda in June, he expressed "sorrow" for colonialism. On a visit to Barbados last year, when the country removed the British monarch as its official head of state, he referred to the "appalling atrocity of slavery."

The prime minister of Barbados, Mia Mottley, called him "a man ahead of his time," in a BBC interview over the weekend. "What has stood out for me is his commitment to the environment, to biodiversity, to urban renewal," Ms. Mottley said.

She is among the most vocal champions of climate action, repeatedly calling on rich countries like Britain and the United States to help repair the damages of climate change to countries like hers.

What Charles has done

As Prince of Wales, Charles spoke out against air pollution, industrial agriculture and deforestation and increasingly called for the world to act on climate change. "The eyes and hopes of the world are upon you to act with all dispatch, and decisively, because time has run out," he told world leaders at the international climate talks in Glasgow in November.

In 2015, Buckingham Palace confirmed that his investments, along with his charitable foundation, include no fossil fuel holdings.

The scheduled appointment was with David Rothschild, whose friendship had endured since their days as schoolboys at Eton.

They greeted warmly as old schoolboy chums often did.

"I truly expected to wait for an appointment with his majesty," said David. "You must have remembered that pass for your try on the rugby pitch."

"The one and only try in my scholastic career," replied Charles. "Actually, what comes to mind was your defense of my position in debate. I have always spoken against the unconscionable seizure of the natural resources of subject nations in colonial India and east Africa. This mistreatment included slavery which continued to mid-nineteenth century. The only bright spots were people like David Livingston, who put slavery under the world's spotlight. To take a controversial position which did not flatter the British Empire required courage and intellectual honesty."

'To severely criticize slavery today takes much less courage than it did two centuries ago', thought David. King Charles was unaware of the activities of the Council of Twelve.

"I personally am supportive of measures to reduce greenhouse gas emissions, but the regent of Great Britain must remain a non-political figure," said King Charles. Some critics have called him out for his use of jumbo jets to ferry a large cortege to meetings and vacation. The jets consume a great deal of aviation fuel. His commitment to building energy grids based on solar and wind technologies is half-hearted, critics allege.

His comments on "overpopulation" in countries of the global south rankled many people, given that the people of those countries have tiny carbon footprints.

His heir, Prince William, heads a conservation group that invests in a fund linked to food companies whose activities contribute to deforestation, The Associated Press reported.

What Britain faces now

Arguably, for the rest of the world, the most consequential climate action that Britain takes now will not be decided by the new king but the new government of Prime Minister Liz Truss.

Truss has said she will ramp up investments in North Sea oil and gas, she has overturned a ban on fracking, and she has chosen Jacob Rees Mogg, one of the few fierce opponents of climate action in British politics, as her new energy minister. He has said he wants the country to extract "every last cubic inch of gas from the North Sea."

It's unclear how this quest will square with a goal enshrined in British law: to cut emissions by 68 percent by 2030, compared to a 1990 baseline. It's the most ambitious climate target of any industrialized nation.

'The British monarch says that he does not engage in politics. So, I don't expect to hear King Charles publicly comment on his country's day-to-day politics,' David thought.

"My son William is attending Climate Week in New York, a series of events this month on the sidelines of the United Nations General Assembly." *Apparently not everyone in the British royal family is apolitical.*

David was aware that that the Crown Estate, which manages a £19 billion portfolio, controls the seabed around the British coastline. That is an increasingly lucrative part of the royal fortune as corporate oil majors seek leases to build offshore wind projects. Industry observers expect the North Sea to sprout tens of thousands of wind turbines if drilling for more oil and natural gas is suspended.

King Charles comments on "overpopulation" in countries of the global south rankled many people, given that the people of those countries have tiny carbon footprints.

His heir, Prince William, heads a conservation group that <u>invests in a fund</u> linked to food companies whose activities contribute to deforestation, The Associated Press reported.

It's unclear how this quest will square with a goal enshrined in British law: to cut emissions by 68 percent by 2030, compared to a 1990 baseline. It's the most ambitious climate target of any industrialized nation.

The British monarch does not engage in politics. David did not expect to hear King Charles publicly comment on his country's day-to-day politics.

It's worth noting though that the Crown Estate, which manages a £19 billion portfolio, controls the seabed around the British coastline. That is an increasingly lucrative part of the royal fortune as oil majors seek leases to build offshore wind projects.

Chapter 25 – **Desperate Measures**

DOW DROPS 3400 POINTS, TRADE SUSPENDED

New York (Associated Press) January 18

In the worst three days of stock market activity in history, the Dow Jones Industrials fell over 3400 points in brisk trading. Although the massive selloff was attributed to a long-awaited market correction, experienced analysts were quick to identify the financial woes of China in the throes of the new SARS-Gamma epidemic . . .

JAPAN CLOSES BORDERS FOR 30 DAYS
TO STEM NEW ASIAN VIRUS

Tokyo (McClatchy News Service) January 19

Officials in the Japanese capital have taken strong steps to quarantine all cases of the SARS-Gamma virus H_3N_7. The government has closed the borders for thirty days to all air and sea traffic. Panicked consumers stripped counters in the stores of a nation which relies heavily on imports. In stocks, the Nikkei Index tumbled . . .

MORTALITY FROM ASIAN VIRUS RISES SHARPLY

by Anthony Wang, Associated Press, Geneva

January 24

Officials in major cities in Africa and India have reported a record number of deaths in a SARS pandemic, Gamma variant. The World Health Organization has described the Gamma virus H_3N_7 as more lethal and more transmissible than the swine flu of 1918-1919 and the corona virus of 2019-2022, which together killed over one hundred million people. Aid organizations have declared a humanitarian disaster. Resources of medical personnel and supplies are exhausted, according to a spokesperson for Doctors Without Borders. The World Health Organization expanded an international travel advisory . . .

London, United Kingdom
January 25 12:00PM

Member Icons

the Terrier   *the Oil Derrick*   *the Crossed Sabers*   *the Top Hat*
the Automobile   *the Cargo Ship*   *the Pyramid*   *the Hammer*
the Rugby Player   *the Candlestick*   *the Hourglass*   *the Falcon*

Members of the Council of Twelve were displayed only as an icon when logging into the video conference. The use of an icon was one method to protect the confidentiality of their identities. Also, the voices of the Council members were modified by a voice synthesizer whenever they spoke. Now, one by one, the icons of the Council members appeared on the large monitor in the conference room.

The conference was the first since October 27 of the previous year. A rift occurred between the Director, David Rothschild, alias 'Q,' and Sir John Clever, alias '*the Terrier*.' The rift turned into a nasty confrontation in which the bombing of the Royal Opera in Covent Garden caused several deaths and extensive damage. The bombing came within seconds of claiming the life of the Director. In retaliation, the security service protecting Rothschild destroyed an expensive yacht owned by Sir John, then employed malware to shut down the corporate business of 'the *Terrier*' costing millions in ransom payments and lost revenue. Four

Belgian mercenaries working for '*the Terrier*' were killed in a street battle outside Rothschild's corporate office.

In the days and weeks succeeding these events, Rothschild and Sir John made peace, each confident that he had made his point. The business of the Council of Twelve was deemed too important to allow egos to overcome common sense, let alone matters concerning the survival of the human race.

After calling the Council to order, *Q* displayed the financial business of the Council including details of the balance sheet – revenues and expenses. *Q* calculated that a spirited discussion would result, a discussion which would highlight his management and focus the activities of the Council on a manageable agenda.

The questions quickly ensued. *The Rugby Player* led off. "The ending balance at our previous conference last year was a little over $252 Million from which we paid $120 Million for the viral epidemic. I see only two other large expenses – the *Lunar Gateway* project for which $50 Million has been disbursed thus far, and $10 Million has been earmarked for increased security. You list 'Revenues' since October last year at over $100 Million which yielded an ending balance of $172 Million. That is amazing! Does the Council participate in some money-making business of which I and other Members are not aware?"

"The by-laws of the Council of Twelve allow the Director to invest in such short-term securities, money markets and liquid exchanges as are suitable to the timeframe that payment of the expenses of Council require," *Q* said. "In 2020, I invested in the Iraqi dinar when that currency was in free-fall during the armed incursion of ISIS. Since the defeat of

168

ISIS in Iraq, the dinar has rebounded nicely. I purchased options to sell the British pound sterling short when a citizen's referendum voted for Brexit-related trade measures. Since then, the pound has tanked. I took short positions on the stock of businesses likely to perform poorly in a recession. The recession has materialized. In all three circumstances, I made informed decisions based of irrefutable logic. Yet I remain subject to your continued confidence in my fiscal management."

"Therefore," said *Q*, "I will take this opportunity to hold a vote of confidence which will be conducted by roll call. Indicate by responding 'aye,' 'nay' or 'abstain'."

Q conducted the roll call quickly. One by one, the Council members all voted 'aye.'

When *the Terrier* affirmed his confidence in the Director, allies among the Council members quickly fell in line.

Next on the schedule was a review by *the Terrier* of the viral epidemic. "The Council of Twelve implements strategies as deemed necessary to achieve the dramatic reduction of Earth's population. The deadly pandemic continues to rage, in the metropolises of Asia and Africa. The connection with the sterilization virus will become apparent in these countries by mid-year."

Q looked to make a show of cordiality and cooperation with *the Terrier* in an effort to demonstrate that animosities were bygone. "*Terrier* - Do you maintain that by midyear the modified HPV will appear and manifest itself significantly?"

The Terrier spoke cautiously.

"I am confident that the designed virus will propagate wildly in Asia, Africa, and South America. Due to the international aid programs available for the treatment of disease, I believe that there will be accurate reporting of the incidence of the deadly virus. The Human Papilloma Virus carried by the SARS/Gamma virus appears as a benign crystalline attachment which is normally ignored in a routine medical diagnosis of illness. The HPV that the Council of Twelve has disseminated is a special virus – *the Crimp* - designed to cause sterility in men. Whether or not the health authorities in various countries choose to report the incidence of HPV and correlate the HPV with a severe drop in pregnancies is another matter."

The Falcon asked, "Are you on schedule to complete this *business* and close the last set of clinics by March 30th?"

"Our program will close sooner since there are several unforeseen complications," replied *the Terrier.* "The spread of this virus is meeting our expectations. However, the publicity generated by the warnings of the World Health Organization has created great concern among the civil administrators who manage the clinics where the virus is inoculated. Prudence warrants those activities of the clinics be curtailed. We will close a month earlier than projected with the last cities being Mexico City, Caracas, Lima, Sao Paulo, and Rio de Janeiro." The timbre of the Terrier's synthesized voice foretold the misery and deaths of untold multitudes in Central and South America.

"When do you think that we will know the complete impact of the epidemic?" asked *the Pyramid.*

"With regard to mortality and sterility?" *the Terrier* responded. "We can obtain a good estimate of global mortality in six to nine months. For the sterility, alarming evidence of plummeting pregnancies will become a matter of crisis for the nations where our clinics have operated. Whether the pandemic achieves the target reduction in world population remains to be seen. The Council of Twelve has made a giant step toward that goal. Duties in this matter terminate after the last clinics close in South America when any trace of the involvement of this Council is eradicated."

"What will be the economic impact of the SARS/Gamma virus and the HPV?" *the Cargo Ship* posed.

"For the nations affected, the economies will suffer due to loss of productivity at work," *Q* said. "Farm production and commerce will be affected. Whenever possible, the prudent will avoid crowded venues and wear a breathing mask. Life in military barracks has been hazardous due to the close proximity of soldiers. Incarcerated prisoners suffer high mortality. Those with financial means will continue to avoid shopping at stores and consumer trade via the internet grows. Tax revenue will suffer. Major purchases for automobiles and homes, however, will not be delayed or cancelled. International trade will decline. Tourism will shrink. A recession will spread worldwide. This is part of the price that must be paid for sustainability. These outcomes are the result of hard but necessary decisions. Whether or not the Council of Twelve should adopt other measures to achieve more drastic reduction in world population is not on the current schedule."

"I present the balance sheet for several purposes."

"Revenues are profits made in financial transactions which involved anticipating how markets respond to adverse conditions. The impact of radical depopulation can be estimated. The future – events determined by the actions of this Council - is assured. Our business interests will succeed by creating opportunities to invest. I urge you to make decisions for your own businesses for marketing conditions that are forthcoming. I will entertain one more question on the matter of the virus."

"Has the Gamma variant virus struck the country of Cote d'Ivoire in Africa?" asked *the Falcon*.

"Let's hear from *the Terrier* for that information."

"Our inoculation clinics *conducted business* in the capital cities of Cote d'Ivoire and its three neighbors Burkina Faso, Ghana, and Liberia. These clinics opened on December 15 and closed last week. The virus should disseminate widely in those countries within a couple weeks."

"Perfect," *the Falcon* commented. Obviously, this information satisfied his need for some relevant business in process.

Q ended the discussion on the viral epidemic.

On the monitor, a display appeared of the International Space Station. A progressive sequence of structural improvements to the old ISS illustrated the projected growth of the project, now dubbed the *Lunar Gateway*. This name recognized the fact that the roles of NASA, the European Space Agency, and other quasi-governmental groups in the development of the project thus named had been coopted by private corporations in the business of space exploration. This situation was a

172

result of an ongoing lack of funding by the U.S. government in NASA which had abandoned the *Lunar Gateway* project in favor of a mission to Mars. Five of the space-exploration companies participating in *Lunar Gateway* were owned by multi-billionaires who were members of the Council of Twelve.

A twenty-frame PowerPoint slide show utilized computer simulation that was absolutely spellbinding. The slides depicted the phases of construction of a virtual city in space. The station was enclosed in a sphere consisting of separate segments of solar panels which when connected become airtight. Within the sphere, an atmosphere and a holographic horizon displayed artificial sunrises, daylight, sunset, and *night*.

Eventually there would be private schools, sports, dining, and entertainment. Gravity would be maintained at 70% of earth's normal gravity. Special medical services and longevity treatments would be available for aging citizens of the *Lunar Gateway*. The city as illustrated boasted attractive *green* habitats and hydroponic farms. A video segment showed machinery which would sustain essential functions of the city, and the last slides stipulated important space industries. The principal industries planned were for the collection and transmission of solar energy and the mining of moon rock – frozen regolith containing ice water, lithium, tritium, deuterium, and helium-3.

"I note that you have leased the International Space Station and moved it to a location deeper into cis lunar space," *the Top Hat* said. "It is widely known that the Chinese recently seeded low-earth orbits with lead pellets as kinetic weapons. These pellets would have been a serious threat to the Space Station. Congratulations on the successful negotiation for the ISS and its timely rescue."

Other Council members added their words of commendation. *Q* basked in the approval that he knew all too well could be fleeting and fickle.

"Allow me to summarize the encrypted report which you may access at the end of this conference," said *Q*. "SpaceX, Boeing Aerospace, Orbital Sciences, Virgin Galactic, Blue Origin, NASA, Northrup Grumman, MAXAR Technologies, and the European Space Agency partner with our proxy identity *Lunar Gateway*. The consortium will assemble on site the components pre-manufactured on earth needed to build an orbital space colony for career astronauts. In the past week, SpaceX boosted the orbit of the ISS from 150 miles above Earth to a Lagrange position 2,750 miles from the Moon. This was quite a project owing to the distance of the journey and the fragility of the old ISS. This event is frame one of the PowerPoint presentations."

The icon of *the Crossed Sabers* flashed. "That moving the ISS that much further away from Earth just makes every step in the construction of the new station – the Lunar Gateway - more expensive. Why was the new site for the *Lunar Gateway* selected?"

"Our decision was based on three important reasons – the balancing of the gravities of the earth and the moon at Lagrange point four, the proximity to the moon, and for defense of the *Lunar Gateway* colony. Let me explain the gravity issue first."

"Newtonian gravitation stipulates that for 'two and one-half bodies,' there exists a point of equilibrium whereby an object of tiny mass would experience equal gravitational attraction. Consider *Lunar Gateway* as that one-half body, an object of tiny mass when compared to the mass of the Earth or the moon. For the Earth and the moon, there are five such points of equilibrium in their elliptical orbits. The average distance of these Lagrange points is three thousand miles from the moon in corresponding orbits. There is significant advantage to locating the *Lunar Gateway* space colony in its present position at equilibrium point L4. *Lunar Gateway* expends little energy required to maintain that orbit. A spectacular advantage to L4 between the Earth and the moon is the existence of a large Trojan asteroid (2010 TK7) in permanent orbit at the L4 location. This asteroid will be a source of water for fuel and the gases necessary for maintenance of life support. As well, there are heavy metals for the necessary radiation shield and silicates for the manufacture of glass."

"Another advantage to the L4 location is its proximity to the moon compared to Low-Earth orbit. At the maximum position, L4 is 2750 miles from the moon. The extraction of resources is key to the economic viability of the *Lunar Gateway.* These resources are primarily - the continual access to the radiation of the sun, secondly H_2O and finally elements like He_3 – helium-3 – and tritium that are rare on earth."

"NASA's Moon Mineralogy Surveyor in 2009 detected water molecules in the lunar landscape and mapped their locations. The H_2O is critical to life support and is also useful for cheap and abundant propellant."

From *the Hourglass* "Where can this water be found? Is it available in substantial concentrations?"

"For some unknown reason, water frozen into the regolith is concentrated in the lunar polar regions. This anomaly is so strange and convenient that in 2012, NASA directed a lunar expedition to gauge precisely the irregular lunar gravitational field. Meteoroids embedded at or near the moon's surface may skew the lunar gravity. The entire surface of the moon is pockmarked from collisions with meteoroids. The impacts from the largest meteoroids have carved out craters. The rims of these craters are quite high and provide Permanently Shaded Areas. PSAs, especially those located at the polar regions, are proven repositories of water in significant quantities."

"Assuming that our prospectors find concentrations of valuable resources, how do we get them to market?" asked *the Cargo Ship*?

"Owing to the Moon's weak and irregular gravity, ore from mining operations on the lunar surface conceivably can be flung from an area of particularly weak gravity."

"Flung?"

"Railguns powered by electricity generated by the sun will fling re-usable nesting tote boxes – containers of ice, helium-3 and rare elements. These tote boxes, full of processed lunar regolith, will be catapulted into an orbit around the moon. The containers can be easily retrieved by robot cargo jockeys for processing at *Lunar Gateway*. This is possible because escape velocity of projectiles from the surface of the moon is just a small fraction of what it is from the surface of Earth. The energy required to put the containers of ore in orbit comes from the sun, free of charge."

"Water is an excellent rocket propellant. There are many scenarios in which a fuel station located at L4 Lagrange would be immensely valuable for spacefaring. For the mission to Mars, refueling at *Lunar Gateway* reduces the cost of achieving escape velocity from Earth by three-quarters and the time of the interplanetary voyage by eighteen months."

"On the matter of self-defense, the time may come when a location of almost 280,000 miles from Earth will make the difference between life or death for the inhabitants of *Lunar Gateway.* The colony may require its own network of defensive lasers and railguns."

"I foresee the excavation of a big hole into which we shovel all our money," predicted *the Oil Derrick*, who obviously resented the demise of *Big Oil.* "Are there any short-term prospects for net return on investment?"

"In the report, there is a segment on a proposal by one of our members for the collection of solar power in space and its transmission by microwave to electric power stations on earth. As we speak, drastic measures are being implemented to reduce man's carbon footprint. More than half of the planet's electricity is generated at power plants which consume mineral resources – coal, oil, and natural gas. Burning these fuels to produce energy results in greenhouse gas emissions. I will pose a question to *the Automobile*: 'Can you summarize the opportunity to sell electric power to utility companies on earth?'"

The response from *The Automobile* was low-key. "Most of the present methods of generating electricity will soon be outlawed, assuming the agenda of the Alliance Manifesto is adopted. The solar collectors and microwave transmitter at *Lunar Gateway* will be automatically deployed and assembled within twelve months. At full capacity, solar power produced and transmitted would generate twice the terawatts used annually on earth. *Lunar Gateway* would have a mostly captive market."

The Oil Derrick and other members heavily invested in the energy sector considered the implications of cheap and abundant solar power.

"For *Lunar Gateway*, imagine a community in space with an almost limitless fuel source generated at location. Visualize a giant artificial habitat for humans to live in and grow their own food with all their energy needs powered by sunlight. People could live on the interior of a hollow cylinder rotating about its axis which creates artificial gravity by centrifugal force. The greater the mass, the more uniform the artificial gravity becomes. Such a colony would increase in mass as the habitat is enlarged with pre-manufactured modules, like those fabricated by Bigelow Environment. Shipping/manufacturing cylinders are attached. A radiation shield, composed of heavy elements mined from the moon, would add considerable mass to *Lunar Gateway*. The extra mass assists in achieving centrifugal force - artificial gravity. Creating earth-like conditions on *Lunar Gateway* is essential for any number of reasons. *Lunar Gateway* could eventually become home to tens of thousands and a popular destination for wealthy tourists. The space station provides a stepping-stone to the stars."

The icon of *the Candlestick* flashed. "You mentioned provisions for the defense of the *Lunar Gateway*."

"The L4 position has a Trojan asteroid containing valuable ore but mostly consists of disposable rock. The asteroid provides two deterrents from attack, both of which employ projectiles possessing kinetic energy. The larger threat posed would be a gravity-assisted bombing making use of big chunks of rock scavenged from the Trojan asteroid. A quicker response with more precise targeting would use smaller projectiles from railguns. Hopefully, the threat of launching kinetic energy strikes should deter aggressors should it become necessary for the defense of the space colony. Let us hope that the *Lunar Gateway* will not have to resort to chucking big rocks at Mother Earth, or for that matter, any moon-based antagonists." Most of those listening presumed that antagonist would be China, operating from a base on the Moon.

"A weapon that provides more precision and versatility is a railgun. Railguns require enormous amounts of pulsed electricity which is not a problem at *Lunar Gateway*. Technical difficulties which dog the effectiveness of a railgun on Earth are more easily accommodated in the cold airlessness of space. A railgun emplacement would deter potential antagonists based in space, on the Earth or coming from the Moon."

The icon of *the Hammer* flashed. "Speaking of kinetic weapons, what threat do the Chinese lead pellets pose to the Artemis rockets that we send up to *Lunar Gateway*?"

Q responded. "The Earth's atmosphere is unforgiving to unshielded objects that fall below an altitude of 120 miles. By week's end, all of the lead pellets will have burned up due to reentry friction."

"When do we start?" *The Terrier* was all aboard for this project. That boded well for making *Lunar Gateway* the number one priority. After the assault on the Royal Opera at Covent Garden, *the Terrier* had been cut off from the flow of information from the Council of Twelve. After this conference, that flow of data would be restored.

"So far we have four payloads delivered to orbit at L4 where the old ISS is now parked. Let me stop the slideshow at stage two which depicts a suited technician in a spacewalk with a wrench in hand. The caption reads: *Deliver initial solar panels, assemble storage batteries and other components for the microwave transmitter, and engage a programmed robot technician to assist in assembly of repetitive tasks*. The fourth payload – stage three - is the delivery of a capsule with two human technicians, life-support supplies, and a space-cargo truck that is called an assembly rover. Work is now in progress for stage four – assembly of an array of solar panels and a microwave transmitter at L4. The astronauts work three-hours shifts with entry and egress from the International Space Station, now dubbed *Lunar Gateway*. Each work session is carefully coordinated and supervised by technicians at our new facility at Kourou."

"Kourou?" *The Cargo Ship* inquired.

"Kourou, French Guiana, has been the spaceport location for the European Space Agency (ESA) since 1975. The global recession made the ESA Spaceport a big financial burden for its European partners. Space-based research used to provide dependable revenue for the Spaceport. Cutbacks in grants and other sources of funding meant a closure of the facility. ESA made an offer to both SpaceX and Orbital Sciences; ESA had no knowledge of the Council or its proxy *Gateway LTD*. The European Space Agency presented a tremendous opportunity to reduce operating expenses and consolidate operations while maintaining anonymity as an affiliate of ESA. Charter bylaws of the Council gave the Director discretionary authority to take advantage of such opportunities. *Gateway LTD*, our corporate entity, has leased most of the ESA Spaceport for the next ten years. That is more than enough time for *Lunar Gateway* to achieve self-sufficiency."

The Rugby Player was fixated on the numbers. "We have financed four launches into Space for only $381 Million?"

"The expense would have been triple +that figure for traditional rocket launches from Hawthorne, California, or Dulles, Virginia, the sites of previous operations. Great savings are available by locating our spaceport at Kourou. Many thanks to *the Rugby Player* - SpaceX recycles all the launch boosters now which reflects considerable savings. The next launch – Artemis IV – scheduled to depart Kourou next month - will contain a crew. Flights with Orion capsules with crew are more expensive. Thereafter, spacecraft with the crew will be delivered into orbit piggyback atop a jumbo Boeing jet. Those crews, flying in Orion spacecraft, will detach from the Boeing jet while in orbit around Earth. The Orion spacecraft will be fully fueled for the final leg – the trip from Earth's orbit to the landing pads constructed on the Moon."

"Gentlemen." *Q* brought the Council to the main topic.

"We must demonstrate to the utility companies a working model of a facility that can generate an almost limitless supply of electric power for Earth. Cheap electric power can usher in an unparalleled age of prosperity and *enrich beyond imagination* the members of this Council."

All of the member icons flashed to discuss a topic near and dear to their hearts.

Encrypted satellite phone conversation

January 25   3PM

The Pyramid: "These developments are shocking. Who could have fore-seen that our oil and gas holdings and our pipeline companies would be so radically threatened by solar power transmitted from an old orbiting space station orbiting in cis-lunar space?"

The Oil Derrick: "We could have assisted 'the Terrier' in his assassination attempt."

The Pyramid: "Spilled milk. If we do not tap into this energy source, the Chinese will. One has to admit that 'Q' is a genius. Would you rather that these events occur without any notice? If we do not prepare immediately for the collapse of fossil fuel-related assets, specifically crude oil & natural gas, we will be left holding the proverbial *hot potato*.

The Oil Derrick: "What are you going to do?"

The Pyramid: "I will take no one else into my confidence. I will act immediately. I will instruct my asset managers to open a hundred securities accounts in New York, Chicago, Tokyo, London, and Zurich. I will disguise my trades with dummy net-zero transactions. Dump my shares on the public stock markets over the next sixty days. Take $500 Billion in short positions on oil futures. Sell my corporations for a dime on the dollar. The market for oil and gas will collapse. Our competitors will assume that we have a scheme to repurchase the corporations on the auction block. Who knows? After the value of shares fall, we will."

The Oil Derrick: "If the price is right."

The oligarchs laughed aloud, oblivious to the fate of millions of workers in the industry and unconcerned with the turmoil that their actions created in world economies.

Encrypted phone conversation
January 25   4PM

James Meredith, alias *The Falcon*, was quite knowledgeable about world trade; he actively traded stocks, currencies, and commodities. Armed with new valuable information which Meredith was confident would affect world market prices for a certain commodity, he called Pete, who brokered trades in the Commodities market. Because of the frequency and size of their transactions, Pete and James were on a first-name basis.

The incoming call went straight to Pete's desk. The caller ID flashed on his monitor. Pete switched the display to video. "Hey, James, what do you know?"

"That is the appropriate question, Pete. At this point, I know just enough about a particular commodity to lose a bundle of cash."

"Let us see if we can end up with our balance sheets in *the black* instead of red ink. What are you considering?"

"Cocoa bean futures. If you can convince me that I am not crazy, I will make a major trade. I want to take a long position. My bet is that the contract value of cocoa beans markedly increases for the next 60 to 180 days."

"That bet would require a lot of courage. We are in a recession," said Pete, "and cocoa beans are harvested year-round. Inventories for the chocolate industries are usually in the form of cocoa butter. What is your incentive to buy?"

"The viral pandemic – Gamma variant - is likely to arrive any day now in West Africa where 97% of the cocoa bean crop is harvested by migrant workers. These migrant workers are mostly child labor, aged ten to fifteen years old."

"I see," said Pete. "James, I didn't know that you carry a social conscience."

Both men laughed.

"I have to ask this question: 'Do you have *insider information*'?"

"No," James lied. "Anybody who can read a newspaper knows about the most recent global viral pandemic. The child migrant work force is vulnerable, and the cocoa bean harvest is dependent on labor that is barely paid a living wage. Also, I have read the Alliance Manifesto. If I am not mistaken, commandment four says in part, 'Laws against child labor will be enacted and enforced.' I think the ban on child labor will gain traction after the virus kills lots of child migrant workers. Will the value of cocoa bean futures increase if the cost of labor significantly increases? I am willing to bet big money that futures rise sharply."

"I like it," said Pete. "Public consumption of chocolate is recession-resistant. Most consumers cannot pass by the grocery shelf displaying chocolates without making a purchase. An increase for the cost of labor may even boost sales if companies like Hershey's and Nestles employ a marketing strategy which promotes the elimination of migrant child labor." Pete was becoming enthused.

"I recently learned that the big chocolate candy producers reduced their working inventories. If child migrant workers were to die in large numbers from the virus, or if the Alliance were to enforce the Manifesto ban on child labor, then cocoa futures may double or triple in value."

"How much are we buying?"

"One hundred million dollars."

"You have that in your Evergreen money market account. Do you want to conduct the transaction at the close of the day?"

"Why wait? I feel like having a Hershey bar."

Both men laughed.

Chapter 26 – **Plague in China**

Assembly Hall, Beijing, China
February 6   2PM

Premier Zhou presided over the emergency meeting of the State Council, its third in three weeks. In attendance were representatives from twenty-two provinces, five autonomous regions, four large municipalities, and two special administrative regions. Four members of the Central Committee observed. At the moment, the attention of these worthy government officials was entirely directed at Dr. Kwan Tseung, chairman of the Central Hospital Authority. Tseung's job was an enviable position were it not for the fact that China was suffering another of a series of epidemics of historic proportions.

"Dr. Tseung," the Premier intoned. "The members of the State Council have read your report. We support your request for additional resources. Tell us now when the Gamma virus will run its course?"

'Good choice of words' thought Tseung. The question was not 'Have you found a cure for the virus?" or 'Have you developed a vaccine that is effective?' The nation's top administrators knew that despite having been given every advantage, Dr. Tseung had no *magic bullet* for the deadly epidemic. Millions had died, and millions more were deathly ill. Hospitals were completely full. Incinerators, some newly constructed, cremated corpses around the clock in an effort to stop the spread of the virus. Many of the sick, turned away from fully occupied hospitals and clinics, died in their homes. Signs mandated by the Central Hospital Authority were nailed to the doors of the sick. Reminiscent of the height of the Black Plague, the local health authorities arranged to collect the bodies for sanitary disposal. Loudspeakers attached atop the cabs of large vans implored survivors to bring out their dead. As a consequence, in many areas, schools and businesses were closed. Assemblies were forbidden. Fields and gardens were left untended. The country was in deep recession and China's currency, the yuan, had lost half its value.

"The SARS/Gamma virus H_3N_7 has proven to be a formidable adversary. You must understand that this contagion has arrived in stages. The first stage of infection, although widespread, was deadly only to the frail, some elderly, and some infants. The first stage lasted 7-10 days, after which, many of those infected seem to have recovered. Then their auto-immune systems failed, and opportunistic disease struck savagely. The most common of these diseases was double lobar pneumonia. Less than one in ten of those infected survive. At the terminal stage, the bodies of the sick turn black from cyanosis. The progression of the disease is interesting. Owing to the source of the virus being bats from the country-side, the Gamma virus migrates from there to the big cities. In the case of this epidemic, the progression has been the reverse. The deadly virus definitely originated in our metropolises. The first phase of the illness has affected over a quarter of our population. The Central Hospital Authority believes that the first phase is nearing its end. Now we must contend with various opportunist diseases which develop among those with weak auto-immune systems."

A member of the Central Committee whispered to Premier Zhou who nodded.

"Dr. Tseung, you mention that *the first wave* of the viral infection is concluding. Should this Council infer that there is a second wave other than the appearance of opportunistic disease?"

"When a bird virus and a bat virus infect the same pig cell, their genes become recombinant, resulting in a new and potentially more lethal virus. I am not clairvoyant; the Authority has no evidence of this occurring as yet. A mutation like Gamma was last seen in the pandemic influenza of 1918-1919 which killed more people than died in all the battles of World War I." Dr. Tseung gathered his strength for the revelation to come.

"This strain of the SARS-Gamma virus H_3N_7 is different than any encountered before. Antibiotics hinder rather than aid the patient's auto-immune response. My research has identified the chemical composition of unusual crystalline structures that adhere to the Gamma variant." Li Tseung made no effort to credit Dr. Xin or Dr. Emily Chou for their discoveries.

"The function of the crystalline structures is clear enough. *The crystals attach to blood proteins, excrete toxins, and convert white blood cells produced by the immune system into weapons that harm the very organs that they are sent to protect.* Furthermore, the crystalline structures each exhibit a periodic pulse."

"What do you deduce from the periodic pulse of each individual crystal?" the Premier inquired.

"My first thought was that this behavior is linked to the growth or the reproductive cycle of the crystals. I detected a minute electrical synapse emitted from the leukocytes and absorbed by the crystals,"

Tseung reported shamelessly. "The analogy I use is like an auto battery accepting a trickle-charge."

"On the basis of my observations, the mutated virus – Gamma - is not the result of random recombination. This mutation is a designed variant construct. Furthermore, the origin of this design is sinister. The Chinese people are under attack."

The State Council erupted in angry pandemonium. In a few minutes, the Premier brought the Council to order. He thanked Dr. Tseung for his report and his conclusions.

As he left the chambers of the State Council, Tseung congratulated himself for deflecting undesirable attention from the pitiful performance of the Central Hospital Authority.

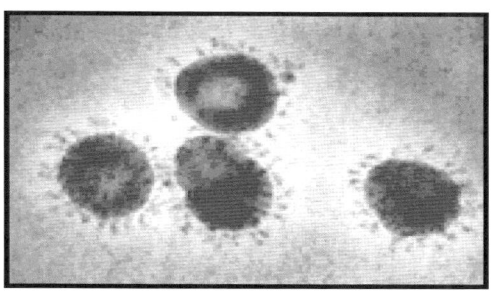

Figure 26. SARS-Covid Images

Human coronavirus 229E virus particles, a coronavirus in the same family as SARS-Covid, as seen in electron microscopic image.

Source: F.A. Murphy and S. Whitfield, Center for Disease Control

Figure 27 SARS-Covid electronic image

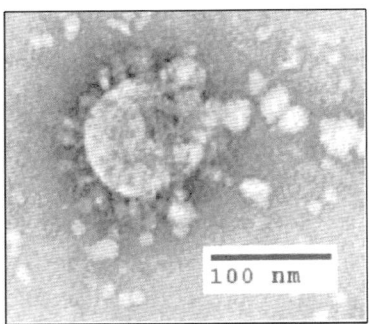

htpps://cdc.gov.sars/lab/images/coronavirus2.gif

Negative stain electron microscopy shows a SARS-Covid particle with club-shaped crystalline surface projections surrounding the periphery of the particle, an unusual feature for coronaviruses. Source: C.D. Humphrey, Center for Disease Control

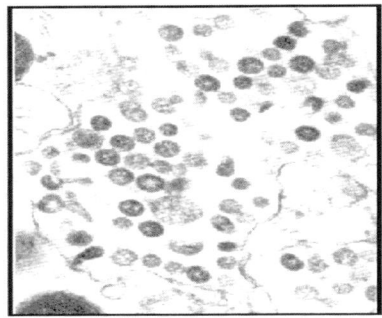

Figure 28. An electron microscopic image of a thin section of SARS-Covid within the cytoplasm of an infected cell, showing the spherical particles and cross-sections through the viral nucleocapsid.

Source: C.S. Goldsmith, Center for Disease Control

Figure 29. SARS-Covid infected cell

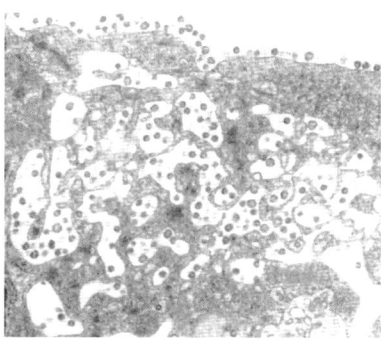

A SARS-Covid-infected cell with virus particles in vesicles, which appear to migrate toward the cell surface and fuse with the plasma membrane, releasing the viral particles. Many of the particles adhere to the plasma membrane, creating an uncharacteristic crystalline-like appearance on the surface of the cell. Source: C.S. Goldsmith, Center for Disease Control

Figure 30 – Utter-chaos-coronavirus-human papiloma virus- exposes-china-healthcare-weaknesses Courtesy of Aljazeera.com

Assembly Hall, Beijing

February 6   3PM

The convocation quickly became a council for war.

The viral epidemic exposed interdependencies that threatened the entire Chinese economy. Widespread employee absence for illness severely cut overall worker productivity which has a widening ripple effect, particularly in key sectors of manufacturing, industry, and transportation. Over the past few decades, inventory of goods had been replaced by frequent deliveries. Most stores and manufacturers had only four days of inventory. Truck drivers failing to show up for essential deliveries of coal caused utility outages which affected households, industry, and commerce. Produce rotted in trucks. Garbage and hazardous waste accumulated. Mass transportation suffered as trains and buses idled for lack of fuel or drivers. In major cities, airlines cancelled more than half of all scheduled flights. Tourism had disappeared. As the epidemic grew worse, employees feared going to their jobs. The epidemic threatened to bring down the whole Chinese economy.

The winds of war had already been fanned by the recent destruction of Chinese military communication satellites in their orbits by SCOLD. Both the United States and China recalled their respective ambassadors for consultation.

The immediate Chinese reaction was a massive cyberattack on U.S. infrastructure – utilities and financial networks mostly. The cyberattack was mostly squashed with some Midwestern U.S. electric grids experiencing short-term outages.

The big question for the Chinese was how to wage war on their best customer, the United States - a customer that, thanks to SCOLD, was now largely insulated against military attack.

The long-term strategy for Chinese hegemony was peaceful coexistence with the U.S.

Winning conflicts was measured in terms of economic output and technological advances.

Prior to the viral epidemic, China had the world's largest economy. The fact that they often resorted to unethical trade practices – government subsidies of exports and theft of intellectual properties, for examples – did not result in warfare. Chinese business interests provided serious competition. The Chinese held over $1 Trillion in U.S. Government Notes and Bonds. Chinese banks possessed over $500 Billion in U.S. currency. China owned large shares in U.S. corporations. A shooting war with the U.S. would devalue these holdings; the U.S. would reluctantly take their business elsewhere. Tourism in China would collapse.

The Premier called for level heads and patience. Colonization of east Africa was proceeding well with Europe and the U.S. unwilling to commit the resources required to contend with the Chinese in Africa. Business opportunities existed in China's energy industry; China was determined to reduce its dependence on coal. China had very profitable business with Canada in the Athabasca Basin which contains the world's second largest reserves of fossil fuel resources. The increased sea level enlarged and deepened the Northwest Passage providing access to those resources as well as offering closer trade routes to the U.S. and Europe.

The Chinese also have plans to exploit energy opportunities and other resources from the moon with outstanding return on investment. The Chinese Space industry is well positioned to follow up recent achievements in missions to the moon.

The Premier allowed speculation that the designed virus may not have been a U.S. conspiracy but may have originated with a third party. Caution was indicated. An ancient Chinese proverb warns that 'man should not cut off nose to spite face.'

CHAPTER 27 - FARSIDE

the Moon, Von Kármán crater

The accomplishments of Cheng'e missions demonstrated new skills adding to the Chinese portfolio of lunar space operations, among which is the first robotic landing on the far side of the moon. In the lunar mission Cheng'e 4, a lander/rover was delivered to a promising lunar location, and Queqiao – a communications relay satellite – braked into the center of a point of gravitational equilibrium known as Lagrange point 2.

The lander/rover, a robotic spacecraft designed to move a two-man crew from location to location, landed in the Von Kármán crater. NASA's Moon Mineralogy Surveyor had detected water molecules in the lunar landscape and mapped their locations in 2009. The Von Kármán crater offered a promising location of impact melt from the southern polar region. The crater, the oldest and largest impact feature on the moon, contained the largest Permanent Shaded Area (PSA). The existence of a sizeable concentration of water was the primary reason that the Chinese selected the site for building a long-term lunar base. Of almost equal importance to that decision were the cliffs of the crater's rim which ranged from 1500 to 1800 feet higher than the surrounding mare surfaces. A solar array could soon be built into the cliffside in position to receive almost continuous sunlight. Energy could be produced for all of the requirements of the planned lunar base; sufficient water was available for life-support and for rocket propellant. Other considerations were for secrecy and defense.

Queqiao, placed in lunar orbit in L2, benefited from a strategic location - one of five spots of Earth-moon gravitational equilibrium. The satellite coasted into this ideal location in December 2018. The Chinese satellite shared L2 with the James Webb Telescope which had operated at that location since May of the previous year. Queqiao functioned as a relay station beyond the moon to link line-of-sight communications with the farside lunar base at the Von Kármán crater. Contingencies were planned for the deployment of a microwave transmitter as a relay for solar energy captured at collectors mounted on the cliffs of the crater.

SCOLD operated from Low-Earth orbits in synchronous positions that were 150 to 180 miles of altitude above Earth. ***Bruised Chinese military egos, frustrated by the dominance of SCOLD, would soon stake a fortress position 240,000 miles away, far beyond the reach of the laser satellites***.

Jiuquan Launch Center
February 7

The *Long March* ballistic booster blasted off its platform without incident, carrying a crew of three and a special habitat module loaded with life-support equipment and supplies. The Cheng'e 6 mission was the sixth stage of the Chinese Lunar Exploration Program (CLEP); the mission was easily the most ambitious by man to explore and establish a human presence on the Moon.

Kingdom of the Air

February 8

Yuan Li and Mephisto were transported to the Simulation Center for a conference that both demons had requested. After arrival, they found themselves attired in the uniforms of infantry soldiers of the 19th century American Civil War. Mephisto wore the dark blue of the Union while Yuan Li was garbed in the light gray of the Confederates. Each sported an era authentic handlebar mustache.

Lucifer appeared moments later in the guise of the Grim Reaper. The adversary of God always had an object lesson for the simulations he employed.

The demons were mystified. Yuan Li was often short-tempered and impatient with Lucifer's intricate schemes. He volunteered to be the obtuse one.

"I thought that the plan was to permit sinful man a pathway to the outer planets and the stars. Yet you have contrived to create a conflict between the Chinese and the Council of Twelve, permitting each a lunar presence bound to lead to armed confrontation. Very shortly, there will again be Mutual Assured Destruction, albeit for the Moon and competing space stations in lunar orbit. The humans will be marooned on a dying planet." Yuan Li's crossed arms and deep frown accentuated his attitude.

Lucifer could not be distracted. "Remind me of the outcome of the War Between the States, beyond the matter of who won, and who lost," the Grim Reaper inquired.

Figure 31. **Battle at Antietam** *courtesy of wearethemighty.com*

"Slavery ended?" Yuan Li guessed.

"Slavery was a dying institution in 1865. By the end of the war, slavery was outlawed in every country except Brazil," Lucifer said. "Let me speed this lesson along."

Lucifer waved his hands and pointed at Yuan Li and Mephisto. Instantly each demon received an antique weapon. Mystified, they examined their weapon in hand, standing about fifty paces apart.

Yuan Li held a muzzle-loaded black powder musket, a weapon commonly carried by Confederate soldiers in the first two years of the Civil War. Sometimes these muskets had rifled barrels; more often the muskets were smooth-bored. The muskets with smooth-bored barrels were wildly inaccurate beyond fifty paces. His musket loaded, Yuan Li took aim at his old foe Mephisto and fired. The mini-ball flew harmlessly astray.

Mephisto was amused. He held a Spencer repeater – a carbine that could be loaded and fired at targets multiple times before a musket could be reloaded. The Spencer was commonly issued to Union soldiers in the

198

first two years of the Civil War. The demon hoisted his weapon and purposely discharged his rifle high in the air – once, twice. "Bang – bang!" Mephisto laughed at Yuan Li's expense. "You're dead!"

The weapons in the hands of the demons transformed. Now Yuan Li held a Spencer carbine. Mephisto, however, possessed a superior weapon, the Springfield rifle. As the crafty demon examined the weapon, a bayonet appeared attached to the end of the barrel.

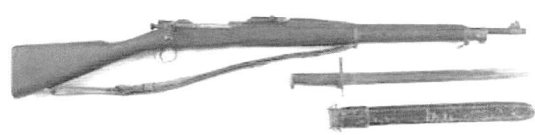

Figure 32. **Springfield Rifle model 1863 .58 caliber with bayonet**

Source: https://teva.contentdm.oclc.org › digital › collection

"ARGH!" Yuan Li cried despondently. "This is not a fair contest. What does it all mean?"

Lucifer explained. "Early in the Civil War, Confederate victories on the battlefield resulted in the capture of thousands of Spencer carbines. Also, southern blockade runners traded cotton for Spencer carbines sold by the British. However, by 1864, the superiority of the Springfield rifle with bayonet was one of a number of factors in the ascendency of the Union in that conflict." Lucifer was a scholar of military history.

"Let me hazard a guess at the meaning of this object lesson," said Mephisto. "Warfare stimulates the advance of technology. Two

Confederate incursions into the North and numerous losses on the battlefield created desperate conditions. Both sides in the conflict looked to technological advances to tip the balance in the war."

"Armored battleships with revolving turrets, submarines, aerial observation, the Gatling Gun – all technologies advanced as the result of the American Civil War," Lucifer declared. "To an even greater extent, technology will advance in a managed conflict between *Lunar Gateway* and *Zhong Guo* colony at the Von Kármán crater."

Both demons were visibly uneasy with this prospect. Without a doubt, Mephisto viewed the human presence at *Lunar Gateway* as fragile and tenuous; Yuan Li had considerable concern for the tiny group of Chinese technicians at *Zhong Guo.* The demons quickly objected.

"The Chinese committed all of its exploration funds to the Moon colony *Zhong Guo,"* said Yuan Li. "No money for a war on the Moon."

"If the Lunar Gateway venture in cis-lunar space fails, the Council of Twelve will dissolve. There will be no real champion for climate control, no powerful advocate for population control, and the likelihood of stalemate in the exploration of space. *God at Judgment will destroy the Kingdom of the Air*. None of God's sinless creations will object. There is no fallback option," Mephisto observed. Lucifer sadly shook his head.

Yuan Li had more input. "The Cheng'e missions were financed when the Chinese economy was robust. The viral epidemic has driven China deep into recession. Resources for future Cheng'e missions cannot be sustained at past or present levels. If *Zhong Guo* fails, no second effort for a lunar colony is foreseeable," concluded Yuan Li.

"Fools!" cried Lucifer. "You have confused the means to a desired outcome with the goal itself." Lucifer was ever the master of strategy. "Tell me, what is our goal for mankind? Do we strive for the extinction of the human race?"

Like scolded school children, the demons acceded to their ultimate authority. "No," Mephisto said. "The primary goal is worship."

Lucifer beamed. "Spiritualism – direct worship of me – is the best outcome and for what the Kingdom of the Air strives. The next best alternative is secular humanism, a godless theology in which man relies on his technical expertise to achieve self-sufficiency. There is no need for God in that formula. Mankind advances to godhood by attendance to its own self-interests. Increased technology feeds those self-interests."

Yuan Li was still unconvinced. "Left to themselves, *Lunar Gateway* and *Zhong Guo* will destroy each other," Yuan Li predicted. "This situation calls for intervention by *the Kingdom of the Air*. Why don't we just encourage spirited athletic competition?"

This time both Lucifer and Mephisto laughed aloud. "The ruling class of every human society hates competition. Surely you are aware that sinful man is motivated by greed, laziness, and fear of failure," Lucifer explained. "Sinful man desires monopoly, which is the natural state of business. Competition only results when two or more parties in a marketing conflict negotiate a resolution which is enforced by the rule of arms."

"Now I pose a question," said Lucifer. "Why don't I simply provide both colonies with force-field technology for protection against attack?"

"That would violate the Rules of Engagement," Mephisto guessed.

"Precisely. Giving the humans *force-field technology* at this time would require a quantum leap of knowledge which is prohibited by the celestial Rules of Engagement. Demons are limited to providing suggestions to humans in step-by-step increments of technology."

"The Chinese and the Council of Twelve realize that their extra-terrestrial outposts may soon become vulnerable to attack from spacecraft using SCOLD-type laser technology. Military outposts on the moon will discourage or destroy invading spacecraft with plasma-operated electro-magnetic railguns. Fear of the destruction of their lunar outposts is a great motivation to discover and harness innovative technologies. This process drives the colonists *away* from God as man becomes more self-reliant."

Lucifer concluded "Sinful man will successfully defend the lunar colonies. *Lunar Gateway* and *Zhong Guo* will each transmit solar power cheaply to Earth, ushering in an era of unparalleled prosperity. Green-house emissions from the consumption of fossil fuels will almost cease, and the Earth will recover. These outcomes result from the acts of man, not God. ***Sinful man will inhabit the outer planets, and eventually journey to the stars. Mankind will extend the limits of his mortality, and <u>all</u> humans will take their place among the children of God.***"

Chapter 28 - **Race for the Moon**

London, United Kingdom

March 18   12:00PM

the Terrier	*the Oil Derrick*	*the Crossed Sabers*
the Top Hat	*the Automobile*	*the Cargo Ship*
the Pyramid	*the Hammer*	*the Rugby Player*
the Candlestick	*the Hourglass*	*the Falcon*

Members of the Council displayed only an icon when logging into the video conference. This conference was extraordinary in that never before had the Council had convened twice in a month. The special meeting was held in recognition of the elevated level of interest in the proceedings – the outcome of the silent auction for shares of ownership in the *Terawatt Project at L4*.

The Council planned to construct a commercial-industrial size solar power system capable of collecting and transmitting energy to generate terawatts of electrical power on Earth.

"The interest among Council members in the *Terawatt Project at L4* is high. The Project requires \$7.2 Billion U.S. The special funding is for the construction of the first array of solar collectors. Since our plan is to develop three or more working assemblies, the *Terawatt Project at L4* will be self-supporting after the first assembly. To conduct our business with all fairness, the Council solicited investment funds by silent auction. Bids were required on March 10; notifications were made on March 11; funds were received yesterday. The investment would have been oversubscribed, so in fairness, each of the members purchased an equal share in the *Terawatt Project at L4*."

"Congratulations to all."

"Launches from Kourou Space Center with the payloads of equipment and personnel will progress with all due haste. The target date for the completion of the *Terawatt* assembly remains May 15 which includes the system check and the first microwave transmission to a corporate power subscriber. In that event, dividends to members will be distributed in mid-July."

The death knell would soon toll for Big Oil, Big Coal, Natural Gas, and other fossil fuels.

Lunar Gateway
March 22

Flush from the success of the solar energy transmission to Kourou Space Center, Captain Dustin Garrett tackled his second mission priority. For the past thirty days, Garrett and a crew member had explored and surveyed Trojan asteroid (2010 TK7), with which *Lunar Gateway* shared at Lagrange point four. The space station was tiny compared to the massive asteroid. *Lunar Gateway* was parked in orbit around the Moon just a scant five miles away from the asteroid.

The most recent launch from Kourou provided the astronauts with an ideal exploration vehicle. Purchased from NASA, the early prototype of the USAF X-37B was perfect for the jobs of scouting the asteroid and transporting rock samples back to the station for analysis. The vehicle would prove to be indispensable to assembling the solar arrays.

The regolith samples contained silicates, bonded with significant quantities of frozen H_2O and rare-earth elements. Surface samples were heavily dusted with helium-3. Yet the helium-3 did not extend deeply into the regolith, indicating an early age for the moon, six to ten thousand years at most. The similarity to the NASA survey of 2009 was an indication of the common origin of the Moon and the captive Trojan asteroid. The crew marveled at the good fortune that the placement of the Moon and the Trojan asteroid was such that they would be of immense value to man's initial explorations into space. Certainly, the Earth is a *privileged planet*.

In their analysis of rock samples, a discovery filled the astronauts with alarm. Every sample showed signs of exposure to solar flares. Garrett engaged the assistance of Earnest Simmons at Mission Control at Kourou Space Center. In preparation for the likelihood of cosmic rays and other high-energy particles penetrating the unprotected *Lunar Gateway*, Garrett and two engineers spent three days nudging the old space station with its new modules closer to the asteroid. Thrusters were mounted with programming to fire automatically to keep *Lunar Gateway* permanently in the protective shadow of the asteroid. For the time being, the Trojan asteroid sheltered man's small colony in space from lethal cosmic rays and other high-energy particles.

A solar flare gives little notice.

Kingdom of the Air

March 1

A Gregorian chant composed itself in the recesses of Yuan Li's consciousness. The demon immediately recognized the music as a telepathic device employed by Lucifer to announce a brief personal conference. In moments, a larger-than-life disembodied face – the face of Lucifer – appeared in the thin air of dark matter.

"Legionnaire." The rich baritone that no human voice could ever replicate spoke clearly.

"Sire." Yuan Li awaited the request – a command – from his master.

"I have good reports of the progress of the viral pandemic and of your cooperation with Legionnaire Mephisto in that matter."

"I exist to serve you, Sire."

"Papal envoys - Jesuits - have been empowered with adequate inducements for the Chinese to accept nuclear disarmament on terms offered by the Alliance. The Chinese must begin inspections of their arsenal and production facilities immediately." The face disappeared, but the sound of the word *immediately* rose to a crescendo and fell in stages to absolute silence.

London, United Kingdom

March 2 12:00PM

Member Icons

the Terrier   the Oil Derrick   the Crossed Sabers   the Top Hat

the Automobile   the Cargo Ship   the Pyramid   the Hammer

the Rugby Player   the Candlestick   the Hourglass   the Falcon

Members of the Council displayed only an icon when logging into the video conference. The use of an icon was one method to protect the confidentiality of their identities. The voices of the Council members were modified by a voice synthesizer whenever they spoke. One by one, the icons of the Council members appeared on the large monitor in the conference room.

The *Terrier* gave his oral report on the progress of the influenza pandemic – a plague designed and orchestrated by the Council of Twelve. This report purported to be the last of a series of reports on the pandemic. Since criminal investigations pursued every rumor, members of the Council had reason for great caution.

"Give us your best estimate of mortalities," Council members demanded.

"Over fifty million deaths directly from the Gamma virus H_3N_7 – and from opportunist disease, another one hundred million. I have no reliable data from Africa, South or Central America." The clinics from which the virus spread closed quickly, in the dead of night in many cases. Clinical staff disappeared, never to be seen again.

"From sources inside the World Health Organization and the Center for Disease Control, males who recover from the virus are often sterile. Females of child-bearing age develop cervical cancer."

Icons flashed.

"The Council has struck a great blow for population control," the *Hourglass* declared.

"The full impact of the virus may never be known," the *Top Hat* surmised. "There may be upwards of 300,000,000 dead and a billion rendered sterile."

The *Terrier* did not refute estimates that were likely too high. Herd immunity would be achieved long before that many fatalities could happen. '*Save that conversation for another time*', he thought. "No fully effective vaccine is available," he said.

"All further discussion on this matter is permanently tabled," the Director stated. "The other business of the Council today are the next steps for *Lunar Gateway*. The first step is to erect a radiation shield to protect the habitat modules of our space station. The alternative is to move the ISS/*Lunar Gateway* close enough to the Trojan asteroid that it shields most of the complex from the dangerous radiation of the sun. We cannot continue to expose our astronauts to cosmic rays and other high-energy solar particles."

"The second step is the construction of a large commercial-industrial size solar power system capable of collecting and transmitting energy to generate terawatts of electrical power on Earth."

"The interest among Council members in the *Terawatt Project at L4* is high. The Project requires $7.2 Billion US. Although our plan is to develop three or more working assemblies, the *Terawatt Project at L4* will be self-supporting after the first assembly. In order to conduct our business with all fairness, the investment funds will be solicited by silent auction. Bids are required on March 10; notification will be made on March 11; funds must be received by March 15. Assuming this investment is fully subscribed in the silent auction, launches from Kourou Space Center with the payloads of equipment and personnel will progress with all due haste."

Several icons flashed. "When will the first assembly of the *Terawatt Project at L4* be completed?"

"Barring unforeseen difficulties, astronauts/ engineers at *Lunar Gateway* will perform final system checks on June 1. Thereafter the *Lunar Gateway* will generate terawatts of electric power to sell to power companies on Earth."

"What future commercial endeavors do you foresee for the *Lunar Gateway*?" asked the 'Crossed Sabers.'

"Our next endeavor will be to develop a world currency based on electric power credits. The Council of Twelve will coopt all money markets and supplant the international value of the U.S. dollar," 'Q' predicted. "Indeed, efforts are underway to equate the major world currencies."

"We provide economic justification to suspend use of fossil fuels – crude oil, natural gas, coal – and advance the crusade for zero greenhouse emissions in the effort to achieve climate control. The Council of Twelve will succeed where world governments have failed."

"Cheap electric power and abundant quantities of helium-3 opens the door for nuclear fusion. We will exploit whatever resources that the Trojan asteroid possesses. We will conduct robotic excavation and processing of regolith containing water, tritium, lithium, and helium-3 from the Moon's north polar region. ***The Lunar Gateway will sell rocket propellant to spacecraft headed for the outer planets and the stars beyond.***"

Chapter 29 - **CONNECTION**

French Guiana
Kourou Space Center
February 24

From his workstation at Kourou, mission control technician Ernest Simmons assisted Captain Dustin Garrett in the final task of the assembly of the microwave transmitter at *Lunar Gateway*. Due to a Zoom camera mounted on astronaut Garrett's head, mission control on Earth enjoyed as good a view as the astronaut. Beyond the spider-web array of solar panels streaming from the space station lay a giant crescent Moon, a smaller Earth, and the cosmos.

The work on this space excursion from the *Lunar Gateway* required the skill and flexibility of a trained human hand. Fine tuning involved the adjustment and calibration of the receiver, the timer, the display remote and the signal processor, all of which had been installed by robotic devices.

Simmons' supervision was particularly helpful as an aid to Garrett's memory and to provide close-up displays on the right lens of bifocals that Garrett wore. The lack of ambient light was the problem. There were three amplitudes of his head lamp, but glare was frequently annoying. With an audible command, the displays on the right lens could provide an angular 3D view which was also useful. The time delay of the communications signal from Earth – a little less than 1.5 seconds - was hardly noticeable.

"That's your last task in assembly, Dustin." During three weeks in work that necessitated constant supervision, the astronauts from *Lunar Gateway* and the technicians from Kourou Mission Control spoke on first-name basis. Three astronauts compiled three-hour shifts totaling 66 hours of extra-module activities in space. The power LED on the remote display was lit in bright green.

"Do you have the power signal, Earnest?" Dustin asked.

"That is affirmative," said the technician from Mission Control. "The director said for you to return to base. The rest of the crew awaits you at the command/communications center for a formal system test at Kourou Space Center. The Space Center has terminated the electric power from the utility company and is operating with electric power from a back-up generator. Henceforth, the Space Center will use solar energy transmitted from *Lunar Gateway*."

Two system checks were conducted – one from Kourou, one from *Lunar Gateway*. The system power-up opened blinds on solar panels that had blocked radiation from the sun. With the blinds opened, immense solar power was tapped. The radiation was converted to microwave energy and transmitted to the receiver at the Space Center. The auxiliary power light went off when an engineer switched off the back-up generator. In a ceremonial gesture to the crew on *Lunar Gateway*, all the lights at the Kourou Space Center blinked off and on.

There was much to celebrate at the Kourou Space Center and at *Lunar Gateway*.

Encrypted notifications sent from the CEOs of SpaceX, Blue Horizon, Northrup Grumman, and Orbital Science brought jubilance to certain members of the Council of Twelve. For other members – *the Oil Derrick* and *the Pyramid* – the news brought increased incentive to dump shares in public stock markets, reduce inventories of crude oil and natural gas by selling on the spot market and take short positions in the world commodities markets. The successful test of a small-scale system opened the avenue for large-scale investment. Microwave transmission of solar radiation from larger arrays collected in cis lunar space by *Lunar Gateway* would soon provide cheap electric power for all of the commercial, industrial, and residential needs of the planet.

This outcome could come at a time of dire need for a world with oceans and an atmosphere permeated with greenhouse gases. *Rational was provided for lawful compliance with the first commandment of the Manifesto – no more use of fossil fuels.*

Zhong Guo base on the Moon
Von Kármán crater
February 25

Shi Huangdi, Wuan Qinjaing and Qianqian Ribao ate, worked, and slept in the inflatable module that was base of operations for *Zhong Guo*. The Chinese technicians had overlapping responsibilities in the monumental task of building living quarters and a functional worksite. The men quickly discovered that comprehensive plans offering step-by-step guidance were overly ambitious and unrealistic; low-level difficulties threatened to become greater emergencies. Unforeseen difficulties added to work schedules which interrupted sleep cycles. The lack of privacy, intended to be short-term, extended through the nineteenth day of the mission. Privacy-related intrusions accentuated irritations to the point of arguments which threatened the chemistry of the team.

Awaiting a daily communication from Beijing, the three Chinese technicians hunkered down in their habitat/ operations module. Each wore a cloth filter/mask over nose and mouth, while two air filters hummed laboriously. Even though an airlock separated the men from the harsh extremes of the Lunar environment, the lunar dust permeated their habitat with every egress and re-entry. The awaited radio signal arrived; each man donned his headset.

"This is Aerospace Center. Hold for Dr. Xiqeoi."

The Lunar astronauts awaited Dr. Duwan Xiqeoi, the chief scientist and director of China's Lunar Exploration Program (CLEP). Almost absent in the headsets was background white-noise which had plagued earlier communications. Ionized dust, suspended in a layer that extended 10-15 feet above the lunar surface, caused interference. The problem had been reduced to the point of a minor annoyance by the assembly of a low-tech hard-wired antenna tower. This task had been achieved after the delay of four days' activities.

"This is Xiqeoi. Joining me in this conference are Cheng'e 6 mission planners, project engineers, and the Aerospace Center physician. Please inform me that you are all safe and in high spirits." This last question was window-dressing. The astronauts knew that Colonel Shi Huangdi, mission commander, reviewed a daily assessment of the physical and emotional health prepared by Qianqian Ribao. Team physician was one of several duties for Dr. Ribao. Xiqeoi was keenly interested in all aspects of the team psyche; these three men had been selected with great care to avoid personal obstacles. These three men represented China's best and brightest, chosen for the best combination of skills, experience, and personalities.

"I will vouch for the health of the crew," offered Dr. Ribao, "if and when Colonel Huangdi will vouch for our safety." The team physician and the Aerospace physician both knew that breathing lunar dust in the habitat and absorbing harmful radiation would be lethal if the current site were maintained much longer.

"The safety of my crew has been the top priority," Huangdi affirmed.

"Although operational problems have occurred, the activities that you have accomplished have provided valuable experience for the current and future missions," Xiqeoi alleged.

The past nineteen days were filled with contingency actions for problems that had arisen at the moon base. These problems had thwarted several early mission objectives. Mission planners were essential for the resolution of whatever difficulties that were encountered. Mission planners were intimately familiar with the equipment and procedures necessary for survival and for mission objectives.

Chief engineer Wuan Qinjaing responded.

"Mission planners have to be disappointed with the lack of progress in survey work. All we have accomplished beside erecting a radio antenna is four trips in the lander/ rover to perimeter points around the crater, where samples of regolith were collected. *I can report that Chinese flags were hoisted at our worksite.*"

Neither the chief scientist for CLEP nor the mission commander spoke.

The engineer continued. "Even the current location of our inflatable module was an alternate site within the Von Kármán crater."

The robotic lander/rover had been delivered to the Far Side at the Von Kármán crater in December 2018. The lander/rover was their primary work vehicle for the survey task of identifying a suitable lava tube (cave) in the cliffside of the crater wherein the moon base would be permanently located.

"Engineer Qinjaing, summarize for Mission Control the reasons for the delay in relocating the habitat/ operations module," said the Colonel.

"Yes. The lander/rover is a robotic spacecraft designed to move a two-man crew for location to location on *the surface* of the Moon. The spacecraft is simply too large and too heavy for entry and egress from the apertures of most of the lava tubes in the rim of the crater. Radar pulses sent by the lander/rover into caves provides reliable information on the size and contours of the lava tubes. The exploration team searches for an empty lava pool close to the top of Von Kármán's rim. Queqiao – a communications relay satellite – orbits the Moon at Lagrange point L2 for

216

the purpose of forwarding line of sight radio transmissions. However, navigating the lander/rover into large lava tubes presents the twin problems of communication signal delay and signal interruption. The desired location for *Zhong Guo* is an empty lava pool with lateral entry by lava tubes, large enough and deep enough to deploy modules for living quarters, a work site, a storage area, and a portal for the lander/rover.

Figure 33. **Lunar lander/rover at the rim of a large crater**

"Program directors for the mission Cheng'e 6 assume that such a cave could be found; if not, *Zhong Guo* would be located temporarily at the base of the highest rim of the crater while robotic excavators bore into the basaltic cliffside to create a domain with the required dimensions." Qinjaing paused for questions.

The chief scientist supplied, "Then I presume that difficulties have emerged for either location."

"Right. Placement of *Zhong Guo* at *the bottom* of the rim meant that access to direct solar power would be eliminated since the site is in a Permanently Shaded Area (PSA). Our alternative source of power, a small radioisotope thermoelectric generator, only provides three hundred watts.

With air filters running continuously in the habitat module, three hundred watts is barely enough for the requirements of life-support."

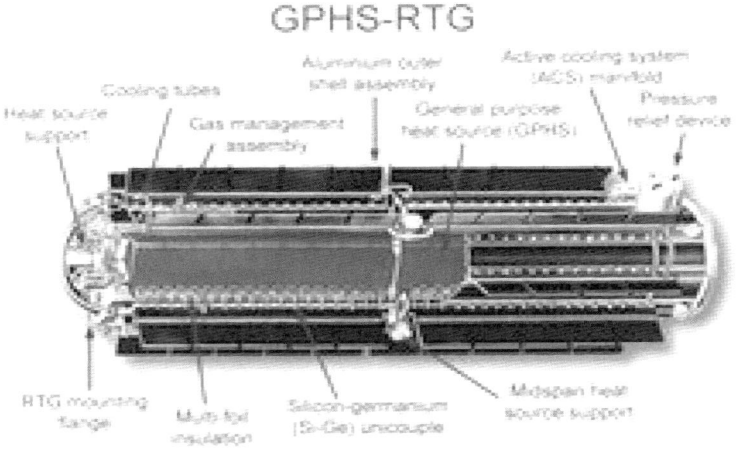

Figure 34. Diagram of an RTG used on the *Cassini* probe
https://en.wikipedia.org/wiki/Radioisotope_thermoelectric_generator

"Placement of *Zhong Guo* at *the top* of the crater's rim means a steep descent of about 1700 feet down into the crater for every task involving the excavation of regolith. Also, problematic is that workers performing tasks in direct sunlight would eventually suffer from the debilitating effects of harmful radiation. Neither location offers a solution for the primary problem, which is lunar dust. Only the lava pool location, with lava tube entry and egress, provides protection against lunar dust and a dense shield against harmful solar radiation."

"Let me speak to these matters *on the record*," interrupted Dr. Ribao who wanted to speak for the record on the issue. "Lunar dust is an insidious threat to both man and machine. About half of lunar regolith is silicate dioxide, prolonged exposure to which causes a deadly disease called silicosis. With regard to harmful solar radiation, each day at our current location increases exposure to cosmic rays and other high-energy particles. The damage to living tissue is cumulative and irreversible."

"The lunar surface is covered with a layer of tiny, sharp-edged crystalline chards – silicates - that are easily disturbed. Small particles become suspended in light gravity; these particles intrude into everything from space suits to moving parts of machinery and even cameras. Our engineers track lunar dust into the lander/ rover and the habitat/ operations module. The lunar dust easily gets into the cabin atmosphere of the lander/ rover; because of lunar dust, astronauts have to wear their suits inside the lander/ rover. As to dust in the module, we cannot afford to allow lunar dust in the habitat atmosphere where it is an irritant to respiratory tracts. It is futile to blow the dust off our space suits outside the module. The exhaust fan in the airlock allows us to blow dust into the air of the airlock where it is vented outside, but this is an unacceptable solution. We estimate that three percent of the module's air is vented with every entry into the habitat."

Dr. Xiqeoi considered this latter problem. The loss of breathable air could not be sustained for long. This dilemma would require action soon. Nonetheless, Xiqeoi tabled further inquiry on the air supply. "I will schedule the silicosis issue high on the agenda of the mission planners."

"Colonel Huangdi, summarize for Mission Control the analysis of your four surveys on the Von Kármán crater."

Colonel Huangdi skippered the lander/ rover and was team leader of surveys. He was quick to respond. "Ice H_2O, helium-3, tritium, lithium, deuterium, and other rare-earth elements are abundant in the regolith found in the Permanently Shadowed Areas of the south polar region. Our surveys within the Von Kármán crater have confirmed the NASA geologic survey taken from low orbit of the Moon. We excavated to a depth of two meters easily because the regolith is loose; it crumbled under our picks and shovels. At the sites of our surveys, about half of the regolith is composed of silicates. Silicates, good for making glass, adhere loosely in the regolith, and are frozen into rock by the presence of ice."

Figure 35. View of the Von Kármán crater from the orbit of the Lunar Gateway

"That should be good news for the mining expedition," the chief scientist said. "Give me your recommendations for excavation."

"Not so fast," said the Colonel. "Shouldn't the placement of our encampment be a higher priority? As you know, there are two types of craters: impact craters and volcanic craters. Among other criteria, the Von Kármán crater was chosen for its volcanic origin. Basalt, a dense igneous rock of volcanic origin, is essential to the question of where to plant *Zhong Guo.* The basalt is at the surface in and along the rim of the crater where our survey team searches for an ancient lava pool. The regolith that we found *within* the Von Kármán crater provides little foundational structure for the moon base. Furthermore, the loose soil and accumulated dust presents problems for driving or flying the lander/rover. When landing, the vehicle disturbs lunar dust; an hour is wasted waiting for the dust to settle enough for members of the survey team to see. On every excursion that we took, the wheels of the lander/rover got stuck in the accumulated dust and loose soil."

"Obviously, a lighter rover with outsized tires has to be designed to operate in loose soil and low gravity," said the engineer. "Extraction of the icy regolith begs for a technological solution. An excavator could easily scoop bucket loads of the frozen soil, laden with gravel-sized chunks of H_2O, He_3 and interspersed with rare-earth elements. But driving back and forth from an excavation site invites trouble. The strip-mining extraction method – using big equipment to dig for water and then haul large chunks of regolith to a processing plant – will not work."

"Any other suggestions, Commander?" asked the chief scientist.

"Engineer Qinjaing and I have debated the cost effectiveness of different methods of extraction. At the risk of losing quantities of H_2O and helium-3, we could melt the silicates and build roadways from excavation site to site, and then to a processing center. There may arise any number of difficulties with this method, including the availability and cost of energy."

"Have you considered pneumatic conveyance of regolith?"

"Pneumatic conveyance of pulverized regolith to a processing center holds possibilities, but with current costs of energy, that method would be expensive."

"How about thermal separation and extraction?"

"Any method of thermal extraction would also require cheap energy because heat transfer from the porous soil *on site* would be extraordinarily low."

"If the crushed regolith could be processed at the deep-freeze of the Permanent Shadowed Area, heating the regolith could distill H_2O and He_3 at different temperatures," the Commander observed. "In the opinions of my team, the pneumatic/distillation process is the best option."

"Data from NASA's Lunar Prospector suggests that over six billion tons of H_2O are sequestered in regolith at the lunar poles. We will engage the most cost-effective plan," the chief scientist avowed, "to capture ice and rare-earth elements, particularly helium-3 and tritium."

In addition to harvesting H_2O, the other motivation for the Chinese mining program is the collection of He_3 in significant quantities. "He_3 (helium-3), a light, nonradioactive isotope, sells for a million dollars an ounce. The Sun has bombarded the Moon with immense quantities of helium-3. Solar wind has embedded He_3 in the upper layers of lunar regolith. Demand for He_3 far exceeds the supply on Earth. Helium-3 would be the perfect fuel, providing safer and cleaner nuclear energy from fusion reactors. Upwards of 1,100,000 tons of He_3 are estimated to exist in regolith on the surface of the Moon."

The chief scientist of Lunar Exploration Program said, "The Moon is so rich in helium-3 that it could meet China's energy demand for millennia and propel our nation to the stars!"

Wuan Qinjaing, chief engineer, was used to speaking his thoughts aloud. Nor was he going to be distracted. "Our outpost here at the Von Kármán is an inefficient use of time and resources. We should have established Zhong Guo as an orbiting space station at Lagrange point L2 where Queqiao – the communications relay satellite – is parked. At L2, an array of solar panels could collect the energy we need and transmit energy *continuously* via microwave to a processing center at Von Kármán. With current equipment and the existing logistical difficulties, construction of a solar array and transmitter on top of the crater's rim is problematic. Why not conduct further exploration surveys and construction of our array from a space station at L2?"

"Then who will guard our territorial flags at Von Kármán?" posed Xiqeoi sarcastically. "Listen carefully. Mission Control has scrubbed all objectives for Cheng'e 6 except two: construction of the solar power station on the crater's rim and selecting a protected site for habitat/operations. The resupply mission - Cheng'e 7 – will land in the crater in 72 hours. A light two-man craft for exploration and assembly is aboard a robotic lander/ ascender. You will supervise the construction of a self-erecting scaffold which will be jettisoned from orbit. The scaffold will support solar panel arrays and transmitters. After the power grid is built and operational, you will resume your search for a cavern – a drained lava pool – for relief from lunar dust and cosmic rays. When you find the suitable location for *Zhong Guo*, then we will rotate the three of you back to China. If you fail, your duty at the Von Kármán crater may become quite lengthy."

The tone of Xiqeoi's instructions indicated definitively that no dissent or further venting of opinions would be tolerated.

"You may download the mission parameters of Cheng'e 7, the recommended excavation procedures, and a complete manifest of the incoming cargo," the chief scientist advised. "The Premier sends his regards and reminds us that the eyes of our nation are upon you."

Chapter 30 – **Negotiations**

Encrypted phone conference
February 28

Pope Luis I: "The Jesuit envoys for the Alliance are meeting resistance from the Chinese to open discussion on nuclear disarmament. The Chinese have a host of complaints."

U.S. President Annette Williams: "Are you aware that the Chinese expelled our ambassador a month ago? That they withdrew their ambassador to our nation 'for consultations'? No wonder why disputes have arisen. That is what ambassadors do – they resolve disputes between nations." (Pause) "What are their specific allegations?"

Pope Luis I: "They say that SCOLD destroyed four communication satellites in orbit. They complain that the James Webb Telescope at the Lagrange point inhibits the function of Queqiao – a communications satellite which relays signals to and from their lunar base. They complain that thousands of Chinese nationals with valid visas have been arrested by the FBI and charged with espionage. They allege that the U.S. targeted and orchestrated a designed virus as a weapon against the Chinese people. Regarding nuclear disarmament, their primary objection – right of national sovereignty - is similar to the complaint of all those nations who possess nuclear weapons. What do you have to say to all these charges?"

President Williams: "It looks like the Chinese handed you their whole dirty laundry list. Allow me to respond briefly."

"The destruction of four Chinese communication satellites was a tit-for-tat response for the destruction of DSS-17, a component of SCOLD, by a Chinese kinetic missile. The U.S. showed considerable restraint in not responding to the destruction of two of its communication satellites."

"Concerning the Webb Telescope at cislunar Lagrange Point 2, that orbiting observatory is an international scientific endeavor organized by NASA whose activities include no military purpose. Webb Telescope offers their research on the universe to everyone. Webb was assembled at L2 well before the arrival of Queqiao. Lagrange point two is large enough to accommodate both Webb and Queqiao."

"Espionage. The U.S. Department of Homeland Security estimated there are more unregistered agents for China operating in the United States than exist for all other nations. They are in the United States for one purpose – to conduct espionage. They steal industrial secrets, they rob intellectual property rights, they co-opt American tech businesses. Through Huawei, Tencent, CCTV and embedded computer chips, they influence public opinion and have even interfered in U.S. general elections. The DHS will continue to weed out foreign agents and their improper activities."

"The SARS-CoVID-2 virus – principally Gamma H_3N_7 – is of unknown origin. If it is a designed construct, the virus did not come from the United States. Attesting to that fact are the Center for Disease Control and the World Health Organization, who have sent teams of researchers at the invitation of China."

Pope Luis I: "All these controversies may provide a providential opportunity to strengthen the authority of our Jesuit envoy to China and provide an example to the rest of the 'nuclear club' of nations. Tell me, what can I offer for bargaining?"

President Williams: "Let me see. I could scale back tariffs, even eliminate them if China decommissions <u>all</u> nuclear weapons. We could ease travel restrictions on Chinese nationals entering the United States. I could drop our objections to Chinese shipping in the Northwest Passage through Canada. I could release 1,000 Chinese spies that are held in detention and expel them back to China. When total nuclear disarmament is achieved in China, you can give them as a reward and a symbol of our cooperation a closely-guarded technology – a small nuclear fission reactor called the Helium Kilo Power System. My Science Advisor tells me that Kilo Power uses helium-3, a non-radioactive isotope. This gift would be conditional to a marketing agreement acceptable to General Dynamics. Finally, as our Jesuit envoys report genuine progress toward complete nuclear disarmament, *the Alliance will permit CLEP, the Chinese exploration and colonization of the Moon, to continue*. The Chinese will not be able to resist these incentives."

Pope Luis I: "I am impressed. Our Jesuit envoys will ensure that there is *quid pro quo*. Every incentive will be held in a tight Jesuit fist."

    The Pope offered a prayer of thanksgiving for the conditions that provided so many opportunities for the growth of His Church.

Kingdom of the Air

March 1

A Gregorian chant composed itself in the recesses of Yuan Li's consciousness. The demon immediately recognized the music as a telepathic device employed by Lucifer to announce a brief personal conference. In moments, a larger-than-life disembodied face – the face of Lucifer – appeared in the thin air of dark matter.

"Legionnaire." The rich baritone that no human voice could ever replicate spoke clearly.

"Sire." Yuan Li awaited the request – a command – from his master.

"I have good reports of the progress of the viral pandemic and of your cooperation with Legionnaire Mephisto in that matter."

"I exist to serve you, Sire."

"Papal envoys - Jesuits - have been empowered with adequate inducements for the Chinese to accept nuclear disarmament on terms offered by the Alliance. The Chinese must begin inspections of their arsenal and production facilities immediately." The face disappeared, but the sound of the word *immediately* rose to a crescendo and fell in stages to absolute silence.

London, United Kingdom

March 2 12:00PM

Member Icons

the Terrier   the Oil Derrick   the Crossed Sabers   the Top Hat

the Automobile   the Cargo Ship   the Pyramid   the Hammer

the Rugby Player   the Candlestick   the Hourglass   the Falcon

Members of the Council displayed only an icon when logging into the video conference. The use of an icon was one method to protect the confidentiality of their identities. The voices of the Council members were modified by a voice synthesizer whenever they spoke. One by one, the icons of the Council members appeared on the large monitor in the conference room.

The *Terrier* gave his oral report on the progress of the viral pandemic – a plague designed and orchestrated by the Council of Twelve. This report purported to be the last of a series of reports on the pandemic. Since criminal investigations pursued every rumor, members of the Council had reason for great caution.

"Give us your best estimate of mortalities," Council members demanded.

"Over fifty million deaths directly from the Gamma virus H_3N_7 – and from opportunist disease, another one hundred million. I have no reliable data from Africa, South or Central America." The clinics from which the virus spread closed quickly, in the dead of night in many cases. Clinical staff disappeared, never to be seen again.

"From sources inside the World Health Organization and the Center for Disease Control, males who recover from the virus are often sterile."

Icons flashed.

"The Council has struck a great blow for *population control*," the *Hourglass* declared.

"The full impact of the virus may never be known," the *Top Hat* surmised. "There may be upwards of 300,000,000 dead and a billion rendered sterile."

The *Terrier* did not refute an estimate that might be too high. Herd immunity would be achieved long before that much mortality could happen. Save that conversation for another time, he thought. "No more vaccine is available," he said.

"All further discussion on this matter is permanently tabled," the Director stated. "The other business of the Council today are the next steps for *Lunar Gateway*. The first step is to erect a radiation shield to protect the habitat modules of our space station. The alternative is to move the ISS/*Lunar Gateway* close enough to the Trojan asteroid that it shields the complex from the sun. We cannot continue to expose our astronauts to cosmic rays and other high-energy solar particles."

"The second step is the construction of a large commercial-industrial size solar power system capable of collecting and transmitting energy to generate terawatts of electrical power on Earth."

"The interest among Council members in the *Terawatt Project at L4* is high. The Project requires $7.2 Billion U.S. Although our plan is to develop three or more working assemblies, the *Terawatt Project at L4* will be self-supporting after the first assembly. In order to conduct our business with all fairness, the investment funds will be solicited by silent auction.

230

Bids are required on March 10; notification will be made on March 11; funds must be received by March 15. Assuming this investment is fully subscribed in the silent auction, launches from Kourou Space Center with the payloads of equipment and personnel will progress with all due haste."

Several icons flashed. "When will the first assembly of the *Terawatt Project at L4* be completed?"

Figure 36. **TERA-WATT PROJECT AT L4**

Courtesy of p.v.magazineusa.com

"Barring unforeseen difficulties, astronauts/ engineers at *Lunar Gateway* will perform final system checks on June 1. Thereafter the *Lunar Gateway* will generate terawatts of electric power to sell to power companies on Earth."

"What future commercial endeavors do you foresee for the *Lunar Gateway*?" asked the 'Crossed Sabers.'

"Our next endeavor will be to develop a world currency based on electric power credits. The Council of Twelve will coopt all money markets and supplant the international value of the U.S. dollar," 'Q' predicted.

"We provide economic justification to suspend use of fossil fuels – crude oil, natural gas, coal – and advance the crusade for zero greenhouse emissions in the effort to achieve climate control. The Council of Twelve will succeed where world governments have failed."

"Cheap electric power and abundant quantities of helium-3 opens the door for nuclear fusion. We will exploit whatever resources that the Trojan asteroid possesses. We will conduct robotic excavation and processing of regolith containing water, tritium, and helium-3 from the Moon's north polar region. *We will sell rocket propellant to spacecraft headed for the outer planets and the stars beyond.*"

Chapter 31 – **Race for the Moon**

London, United Kingdom

March 18   12:00PM

Member Icons

the Terrier	*the Oil Derrick*	*the Crossed Sabers*
the Top Hat	*the Automobile*	*the Cargo Ship*
the Pyramid	*the Hammer*	*the Rugby Player*
the Candlestick	*the Hourglass*	*the Falcon*

Members of the Council displayed only an icon when logging into the video conference. This conference was extraordinary in that never before had the Council had convened twice in a month. The special meeting was held in recognition of the elevated level of interest in the proceedings – the outcome of the silent auction for shares of ownership in the *Terawatt Project at L4*.

The Council planned to construct a commercial-industrial size solar power system capable of collecting and transmitting energy to generate terawatts of electrical power on Earth.

"The interest among Council members in the *Terawatt Project at L4* is high. The Project requires $7.2 Billion U.S. The special funding is for the construction of the first array of solar collectors. Since our plan is to develop three or more working assemblies, the *Terawatt Project at L4* will be self-supporting after the first assembly.

"Launches from Kourou Space Center with the payloads of equipment and personnel will progress with all due haste. The target date for the completion of the *Terawatt* assembly remains May 15 which includes the system check and the first microwave transmission to a corporate power subscriber. In that event, dividends to members will be distributed in mid-July."

The death knell would soon toll for *Big Oil, Big Coal, and other fossil fuels.*

Lunar Gateway
March 22

Flush from the success of the solar energy transmission to Kourou Space Center, Captain Dustin Garrett tackled his second mission priority. For the past thirty days, Garrett and a crew member had explored and surveyed Trojan asteroid (2010 TK7), with which *Lunar Gateway* shared at Lagrange point four. The space station was tiny compared to the massive asteroid. *Lunar Gateway* was parked in orbit around the Moon just a scant twelve kilometers away from the asteroid.

The most recent launch from Kourou provided the astronauts with an ideal exploration vehicle. Purchased from NASA, the early prototype of the USAF X-37B was perfect for the jobs of scouting the asteroid and transporting rock samples back to the station for analysis. The vehicle would prove to be indispensable to assembling the solar arrays.

The regolith samples contained silicates, bonded with significant quantities of frozen H_2O. Surface samples were heavily dusted with helium-3. Yet the helium-3 did not extend deeply into the regolith, indicating an early age for the moon, six to ten thousand years at most. The similarity to the NASA survey of 2009 was an indication of the common origin of the Moon and the captive Trojan asteroid. The crew marveled at the good fortune that the placement of the Moon and the Trojan asteroid was such that they would be of immense value to man's initial explorations into space. Certainly, the Earth is a *privileged planet*.

In their analysis of rock samples, a discovery filled the astronauts with alarm. Every sample showed signs of exposure to solar flares. Garrett engaged the assistance of Earnest Simmons at Mission Control at Kourou Space Center. In preparation for the likelihood of cosmic rays and other high-energy particles penetrating the unprotected *Lunar Gateway*, Garrett and two engineers spent three days nudging the old space station with its new modules closer to the asteroid. The boosters were programmed to fire automatically to keep *Lunar Gateway* permanently in the protective shadow of the asteroid. For the time being, the Trojan asteroid sheltered man's small colony in space from lethal cosmic rays and other high-energy particles.

A solar flare gives little notice.

Kingdom of the Air

March 1

A Gregorian chant composed itself in the recesses of Yuan Li's consciousness. The demon immediately recognized the music as a telepathic device employed by Lucifer to announce a brief personal conference. In moments, a larger-than-life disembodied face – the face of Lucifer – appeared in the thin air of dark matter.

"Legionnaire."

The rich baritone that no human voice could ever replicate spoke clearly.

"Sire." Yuan Li awaited the request – a command – from his master.

"I have good reports of the progress of the viral pandemic and of your cooperation with Legionnaire Mephisto in that matter."

"I exist to serve you, Sire."

"Papal envoys - Jesuits - have been empowered with adequate inducements for the Chinese to accept nuclear disarmament on terms offered by the Alliance. The Chinese must begin inspections of their arsenal and production facilities immediately." The face disappeared, but the sound of the word *immediately* rose to a crescendo and fell in stages to absolute silence.

London, United Kingdom

March 2 12:00PM

Member Icons

the Terrier   the Oil Derrick   the Crossed Sabers   the Top Hat
the Automobile   the Cargo Ship   the Pyramid   the Hammer
the Rugby Player   the Candlestick   the Hourglass   the Falcon

Members of the Council displayed only an icon when logging into the video conference. The use of an icon was one method to protect the confidentiality of their identities. The voices of the Council members were modified by a voice synthesizer whenever they spoke. One by one, the icons of the Council members appeared on the large monitor in the conference room.

The *Terrier* gave his oral report on the progress of the influenza pandemic – a plague designed and orchestrated by the Council of Twelve. This report purported to be the last of a series of reports on the pandemic. Since criminal investigations pursued every rumor, members of the Council had reason for great caution.

"Give us your best estimate of mortalities," Council members demanded.

"Over fifty million deaths directly from the SARS-Gamma virus H_3N_7 – and from opportunist disease, another one hundred million. I have no reliable data from Africa, South or Central America." The clinics from which the virus spread closed quickly, in the dead of night in many cases. Clinical staff disappeared, never to be seen again.

"From sources inside the World Health Organization and the Center for Disease Control, males who recover from the virus are often sterile."

Icons flashed.

"The Council has struck a great blow for population control," the *Hourglass* declared.

"The full impact of the virus may never be known," the *Top Hat* surmised. "There may be upwards of 300,000,000 dead and a billion rendered sterile."

The *Terrier* did not refute an estimate that might be too high. Herd immunity would be achieved long before that much mortality could happen. Save that conversation for another time, he thought. "No more vaccine is available," he said.

"All further discussion on this matter is permanently tabled," the Director stated. "The other business of the Council today are the next steps for *Lunar Gateway*. The first step is to erect a radiation shield to protect the habitat modules of our space station. The alternative is to move the ISS/*Lunar Gateway* close enough to the Trojan asteroid that it shields most of the complex from the dangerous radiation of the sun. We cannot continue to expose our astronauts to cosmic rays and other high-energy solar particles."

"The second step is the construction of a large commercial-industrial size solar power system capable of collecting and transmitting energy to generate terawatts of electrical power on Earth."

"The interest among Council members in the *Terawatt Project at L4* is high. The Project requires $7.2 Billion US. Although our plan is to develop three or more working assemblies, the *Terawatt Project at L4* will be self-supporting after the first assembly. In order to conduct our business with all fairness, the investment funds will be solicited by silent auction. Bids are required on March 10; notification will be made on March 11; funds must be received by March 15. Assuming this investment is fully subscribed in the silent auction, launches from Kourou Space Center with the payloads of equipment and personnel will progress with all due haste."

Several icons flashed. "When will the first assembly of the *Terawatt Project at L4* be completed?"

"Barring unforeseen difficulties, astronauts/ engineers at *Lunar Gateway* will perform final system checks on June 1. Thereafter the *Lunar Gateway* will generate terawatts of electric power to sell to power companies on Earth."

"What future commercial endeavors do you foresee for the *Lunar Gateway*?" asked the 'Crossed Sabers.'

"Our next endeavor will be to develop a world currency based on electric power credits. The Council of Twelve will coopt all money markets and supplant the international value of the U.S. dollar," 'Q' predicted. "Indeed, efforts are underway to equate the two currencies."

"We provide economic justification to suspend use of fossil fuels – crude oil, natural gas, coal – and advance the crusade for zero greenhouse emissions in the effort to achieve climate control. **The Council of Twelve will succeed where world governments have failed.**"

"Cheap electric power and abundant quantities of helium-3 opens the door for nuclear fusion. We will exploit whatever resources that the Trojan asteroid possesses. We will conduct robotic excavation and processing of regolith containing water, tritium, and helium-3 from the Moon's north polar region. *We will sell rocket propellant to spacecraft headed for the outer planets and the stars beyond.*"

Chapter 32 – **Progress?**

London, United Kingdom

March 18   12:00PM

Member Icons

the Terrier	*the Oil Derrick*	*the Crossed Sabers*
the Top Hat	*the Automobile*	*the Cargo Ship*
the Pyramid	*the Hammer*	*the Rugby Player*
the Candlestick	*the Hourglass*	*the Falcon*

Members of the Council displayed only an icon when logging into the video conference. This conference was extraordinary in that never before had the Council had convened twice in a month. The special meeting was held in recognition of the elevated level of interest in the proceedings – the outcome of the silent auction for shares of ownership in the *Terawatt Project at L4*.

The Council planned to construct a commercial-industrial size solar power system capable of collecting and transmitting energy to generate terawatts of electrical power on Earth.

"The interest among Council members in the *Terawatt Project at L4* is high. The Project requires \$7.2 Billion US. The special funding is for the construction of the first array of solar collectors. Since our plan is to develop three or more working assemblies, the *Terawatt Project at L4* will be self-supporting after the first assembly. To conduct our business with all fairness, the Council solicited investment funds by silent auction. Bids were required on March 10; notifications were made on March 11; funds were received yesterday. The investment would have been oversubscribed, so in fairness, each of the members purchased an equal share in the *Terawatt Project at L4*."

"Congratulations to all."

"Launches from Kourou Space Center with the payloads of equipment and personnel will progress with all due haste. The target date for the completion of the *Terawatt* assembly remains May 15 which includes the system check and the first microwave transmission to a corporate power subscriber. In that event, dividends to members will be distributed in mid-July."

The death knell would soon toll for fossil fuel consumption – especially *Big Oil* and *Big Coal*.

Lunar Gateway
March 22

Flush from the success of the solar energy transmission to Kourou Space Center, Captain Dustin Garrett tackled his second mission priority. For the past thirty days, Garrett and a crew member had explored and surveyed Trojan asteroid (2010 TK7), with which *Lunar Gateway* shared their position at Lagrange point four. The space station was tiny compared to the massive asteroid. *Lunar Gateway* was parked in orbit around the Moon just a scant five miles away from the huge asteroid.

The most recent launch from Kourou provided the astronauts with an ideal exploration vehicle. Purchased from NASA, the early prototype of the USAF X-37B was perfect for the jobs of scouting the asteroid and transporting rock samples back to the station for analysis. The vehicle proved to be indispensable to assembling the solar arrays.

The regolith samples contained silicates, bonded with significant quantities of frozen H_2O. Surface samples were heavily dusted with helium-3. The similarity to the NASA survey of 2009 was an indication of the common origin of the Moon and the captive Trojan asteroid. The crew marveled at the good fortune that the placement of the asteroid was such that it would be of immense value to man's initial explorations. Certainly, the men in space recognized Earth as a *privileged planet*.

In their analysis of rock samples, a discovery filled the astronauts with alarm. Every sample showed signs of exposure to solar flares. Garrett engaged the assistance of Earnest Simmons at Mission Control at Kourou Space Center. In preparation for the likelihood of cosmic rays and other high-energy particles penetrating the unprotected *Lunar Gateway*, Garrett and two engineers spent three days nudging the old ISS space station with its new modules closer to the asteroid. Boosters were programmed to fire automatically to keep *Lunar Gateway* permanently in the shadow of the asteroid. For the time being, the Trojan asteroid sheltered man's small presence in space from lethal cosmic rays and other high-energy particles.

A solar flare gives little notice.

London, United Kingdom
March 22

Captain Charles Wilshire of Avent Security completed the routine assessment, which ended at David Rothschild's spacious office. The

Director had his back to the entrance, studying a model skyline of the business section off London.

"I haven't read of any of your activities in *The Daily Mail*," said Wilshire, hoping to elicit a reaction.

"If you are referring to the activities of the Council of Twelve, my response is 'so far, so good.' None of the Council members have breached their oaths of secrecy. Our agenda to curb the growth of the world population is undetected. Our affiliation with Kourou Spaceport and *Lunar Gateway* is hidden from public records." David slowly swiveled in the leather chair, now facing the high-price security contractor. "If you refer to my personal affairs, owing to increased responsibilities and *the protocols adopted for personal security*, I rarely leave the luxurious confines of the New Horizon Towers."

"I am glad that you purchased the entire building including the high-rise condominiums. The Agency has completed the immediate improvements undertaken for purposes of security," Wilshire said.

"The primary purpose for the acquisition of the building is to showcase new realities for urban survival during global climate change," David advised. "Due to rapidly melting ice caps at the polar regions, the River Thames is swollen from increased rainfall and rising sea levels. Soon the river will reach consistent levels of fifteen feet above flood stage. Buildings along the river and at lower elevations have to be demolished or completely mitigated. Buildings have to be undergirded with stronger foundations and strengthened to withstand near–hurricane winds."

"Here, let me show you." David crossed his office over to the scale model of a building he had designed with the assistance of an architect.

"This model is a concept building serving as an example of what can survive and function during the gathering storms and rising river. Much of London will become half-submerged like Venice unless radical solutions are adopted," David predicted.

"These radical solutions come at a price. Most of the mitigations for buildings and homes, such as shoring up and raising foundations, are expensive, well beyond the means of the average homeowner or landlord."

David planned to lean heavily upon modern technologies. "*New Horizon Towers*, as I have renamed this acquisition, is not only the tallest building in the financial district of London but will be deemed the safest. I have approved the assembly of a vertical isolation system that separates the building from the ground using more than 1,000 huge shock absorbers. The system works cooperatively with lateral isolation dampers consisting of huge sliding bearings. The isolation/damper system, a massive-scale version of a luxury car suspension, would be effective against hurricane-force winds or earthquakes."

"That will cost a pretty penny," Wilshire guessed.

"Over one hundred-million-pound sterling," David admitted.

"Let me share some Council investment plans which are at the initial stage." David inserted a flash drive into a wall computer/monitor. A slide show was presented. "A partnership venture with Elon Musk, an innovative corporate billionaire, is promising. Wind turbines in the North Sea are under construction. The turbines generate power which could transmit to satellites which would beam the power to *New Horizon Towers* and other subscribers."

"Incidentally, Musk has developed a currency composed of kilowatt energy credits."

"Sounds to me like the rich getting richer," Wilshire commented.

"There should always be a reward for innovation and risk in a capital-based economy." David spoke from a core conviction. "Especially when that risk is taken in defense of the environment."

"Do you foresee any problems for your *New Horizon Towers* concept?" Wilshire asked.

"Operationally or external hazards? Clean water, water pressure, and sewage treatment are important considerations for the *New Horizon Towers*. The chief external hazard is *perpetual war*."

For that observation, Wilshire had no comment.

"Access and egress from the New Horizon will be by boat, helicopter or gyroplane," David proposed. "Elon Musk has partnered with Germany's largest gyroplane manufacturer to produce a four-seat short-take off-or-landing (STOL) aircraft that could arrive and depart from the top of the building. I will finance the $250,000 gyroplane for all qualified tenants."

"Look at the buildings that surround New Horizon Towers." The slide show presented a GPS overhead map of London's financial district. "I have acquired adjacent properties at depressed prices. My goal is to construct a block of inter-connected buildings with a design capacity for 1,800 condominiums. Upper floors are reserved for restaurants, a fitness center, a gift shop, offices, and my penthouse. Wilshire inspected the scale model, which sat on a table.

Captain Wilshire was finishing a weekly security assessment of the upper floors of the New Horizon Towers. Five months ago, Avent Security and G45 joined forces to thwart an attack on David Rothschild authorized by Sir John Clever, a/k/a the *Terrier* of the Council of Twelve. To demonstrate his pique at Rothschild over matters of Council leadership, Sir John orchestrated a bombing which occurred at The Opera House at Covent Garten. The bomb narrowly missed killing Rothschild. The attack resulted in counter strikes including an assault on corporate cyber networks, the destruction of Sir John's expensive yacht, and the execution of Sir John's Belgian mercenaries. At Sir John's behest, a board member of Barclays Bank hastily brokered a truce. The business deemed critical to the survival of mankind resumed.

Wilshire accomplished his duties and made a roguish inquiry. "I presume that Sir John is in good health?"

David snorted. "Sir John and I are not friends."

"It is my responsibility to monitor the state of your relationship with Sir John. If I am remiss in this regard, you two school children may have a tiff." A finger from the Wilshire's big hand was levelled in Rothschild's direction. "From past experience, I observe that when you and Sir John argue, people die. With the aid of Morgan Trussler at G45, I barely squashed a formal inquiry by Scotland Yard into the bombing at the Opera House."

"Besides that, you pay the Agency a lot of money to protect you." Wilshire smiled broadly. "Asking uncomfortable questions is a part of my job. For example, do you ever entertain any ladies of the night?" asked Wilshire with a wink.

"Go away."

"I depart with a word of advice," said Captain Wilshire. "Have you heard the old adage that 'it is wise to keep your friends close and your enemies closer?"

"Yes."

As Wilshire departed, he remarked, "I would take that advice with a grain of salt."

As if on cue, Rothschild's personal secretary spoke on his phone intercom. "There is a gentleman from Barclays Bank who will not provide his name or leave a message. He said that you would take his call."

Captain Wilshire laughed. Obviously, Sir John was calling – an enemy pretending now to be David's friend.

Zhong Guo colony at the Von Kármán crater
March 23

As scheduled, the resupply mission - Cheng'e 7 – arrived timely, landing in the crater nearby the habitat/operations module. The ascender/lander carried much needed supplies, five greenhouse habitats and a light two-man craft. The craft was little more than an erector-set box frame with swiveled thrusters, useful for assembly operations in the light Moon gravity.

A self-erecting scaffold for support of the array of solar collectors was jettisoned on March 2 by the Cheng'e 7 rocket in its orbit around the Moon. Supported by balloons, the scaffold descended successfully to the desired location atop the rim of the Von Kármán crater.

The Commander and the engineer rapidly gained skills in the use of their new thruster-powered flyer. Using jacks, winches, and pry bars, the men tugged and pulled the collapsible scaffold into a level position on a firm rock foundation at the rim of the crater. Thereafter the Chinese technicians made an electrical connection and moved a lever to telescope the scaffold upward. The men pursued their task one level at a time. With each level, two arms for the solar collectors raised. On each arm, fifty giant solar collectors were affixed and connected to the central power relay. Each time a level was outfitted with its allotted collectors, the lever was moved to the next upward position which exposed two more arms. Over the previous three weeks, seven levels of fifty solar collectors were completed. One more level remained as did attachment of an electric rotational motor and the microwave transmitter. By March 23, the primary objective of Commander Shi Huangdi and engineer Wuan Qinjaing – the construction of the solar power station on the crater's rim – was well on the way to completion. The power grid would soon be assembled and operational. This work would only require a few more three-hour shifts.

The second objective was the selection of a protected site for habitat/ operations - a permanent location for the *Zhong Guo* colony. Once each 24 hours, for three hours, Dr. Ribao resumed the search for a cavern – a drained lava pool – for relief from lunar dust and cosmic rays. From his experience of previous searches, he had determined which parts of the crater's rim were the least illuminated during the Moon's orbit.

While Commander Shi Huangdi and engineer Wuan Qinjiang completed critical mission objectives, Dr. Ribao's efforts were no less important to the sustenance of the fragile presence of the Chinese in the

248

Von Kármán crater. As physician and botanist for Cheng'e 6 & 7, Ribao assembled a self-contained greenhouse – an enclosed habitat surrounded by the airless environs of the Moon. He inflated one of five new habitat modules which had been delivered by Cheng'e 7. Compressed air released from 200 lb. cylindrical tanks, the same as carried by the astronauts, inflated the greenhouse habitat. The 20-meter by 40-meter module was reserved for growing rice and soybeans. The plants received nutrition by composting organic waste, as available.

A thick low-density clear PVC ceiling permitted entry of the sun's radiation and held temperatures of the lunar day. The current site was outside the Permanently Shaded Area and in position to receive eighty percent of the solar constant. Six one-meter-square solar panels provided power for the habitat's heater. Refinements to the array under construction at the crater's rim could increase the sunlight that the botanical habitat received.

Other botanical supplies, a part of the capsule payload of Cheng'e 7, served a growing part of the needs of the astronauts for food and breathable air.

The Commander and the Engineer returned from the crater's rim where they were erecting the array of solar panels on the support scaffolding. After working for three hours on the array, the astronauts had to wait an hour in the airlock while the filters removed tiny shards of silicates. Located on each side of the airlock and the entry to the habitat module was an intercom. The crew soon adapted their schedule to accommodate what would have been inactive time. Sitting on makeshift stools placed at either side of the airlock, the Commander, the Engineer, and the Physician/botanist/lunar cave explorer used the time productively by sharing reports on their individual progress. Usually, the Engineer was detailed in his report, while the Commander's report was short and summarized. Each embraced their tasks with enthusiasm.

Commander Huangdi expressed curiosity about the importance of the botanical module. "Why the emphasis on the early start to a botanical habitat?"

"There are so many ways to die on the Moon," said Dr. Ribao. "The value of the botanical habitat helps us avoid two of these ways: starving to death and choking on our own fumes. Let me pose a question or two." The physician leaned so close that their helmets almost touched.

"As engineers selected for this mission, you know that there is a CO_2 scrubber that vents all this potentially lethal gas out of the airlock every time that the airlock closes. Did you know that we vent oxygen and nitrogen as well? You know that we have a limited supply of compressed air."

"The mission planners of Cheng's 7 shipped us thousands of tiny seedlings which provide a long-term solution to this dilemma. These plants – tiny ferns – have a unique ability to capture carbon dioxide and nitrogen from our sewage and the air in our habitat modules. This fern is called Azolla. Since the dawn of agriculture, subsistence farmers in southeast Asia have deliberately cultivated Azolla as a companion plant for rice. Azolla is a floating fern that will thrive in our farm habitat, fixing nitrogen and other nutrients, constantly improving the soil composition, and providing a natural green fertilizer that significantly bolsters rice and soybean productivity."

Dr. Ribao continued. "The secret here is that Azolla is not just a plant; it is a 'superorganism,' a symbiotic collaboration of a plant and a powerful microbe. In a special protective cavity inside each leaf of this tiny fern, Azolla hosts a microbe called Nostoc. Nostoc spends its entire life converting nitrogen into food for its host. We are talking about a plant that can double its entire body mass in less than two days. Azolla and Nostoc have clearly demonstrated an immense ability to convert nitrogen and carbon dioxide into rice and soybeans."[i]

"So eventually the CO_2 scrubber becomes a redundant system," said Huangdi.

"I will monitor its function; when we no longer need the scrubber, it will shut off by itself and thereby reduce the load on our generator," added Qinjaing.

"How about the airlock filters, aren't they redundant also?" Huangdi probed. "The airlock filters remove most of the lunar dust in the first fifteen minutes after our entry." The Commander had decided that in order to save time, the crew returning from the activities outside the module would not wait for the full hour that the airlock air filters required to remove all of the lunar dust. His reasoning was the air filters inside the habitat module was doing an adequate job in fifteen minutes.

Ribao listened with alarm. It was apparent that the Commander has made up his mind; Huangdi had already conferred with Dr. Xiqeoi, the Director. Jiuguan mission planners had already approved the reduction in air filter use after each entry. Wasn't the air filter in the airlock operating at full capacity for an hour after each *departure*?

Qinjaing looked uncomfortable with this decision. He seldom questioned the Commander's decisions.

Still, Ribao had reason to be disturbed. This command decision encroached upon Ribao's authority as mission physician. "How about if we conduct a little experiment?" Ribao offered. "We examine the lunar dust collected on the filter in this one-hour session and compare what we collect in fifteen minutes after the next airlock entry? I will take lunar dust measurements and photos of accumulated chards."

It was a reasonable proposal, but Huangdi's response made it clear that the order to reduce the wait in the airlock originated with Mission Control.

"A reduction of 45 minutes in the airlock with each entry means that over the next three earth-days we retain enough air in our tanks for eighteen additional hours at the work site. Engineer Qinjaing and I are at a critical stage in the assembly of the solar array." That assertion left little room for argument.

Dr. Ribao dropped his objection. "For the next three earth-days, I will split my work duties between the botanical habitat and searching for a lava pool near the crater's rim. I will move my bunk roll to the botanical habitat. There is no need to expose three men unnecessarily to *silicosis*."

"With any luck, I will find a suitable location to move our operations habitat module out of harm's way from solar cosmic rays. I will be available at all times on the intercom or on the radio communications channel."

Self-preservation was Dr. Ribao's primary responsibility, and to prevent the needless sacrifice of the men with whom he served. Dr. Ribao was greatly concerned that the Captain and the engineer would be exposed unnecessarily to silicon dioxide in the air they breathed or lethal solar radiation that could penetrate their habitat.

CHAPTER 33 - **Danger**

Rim of the Von Kármán crater

March 25   1100 Hours Sol

The southeastern rim blocked all sun light into a wide section of the Permanently Shaded Area, resulting in creating as dark and cold a place as any in the universe. It was at such a location on Dr. Ribao's third excursion that the physician/explorer found a good site for the *Zhong Guo* colony.

The empty lava pool ended abruptly at cliffside of the crater. Ribao set the lander/rover at hover as he explored the interior with his laser camera. The aperture provided ample room for the lander/rover to enter and land. The entrance of an immense cave was spacious and level.

The lander/rover hovered momentarily and landed firmly on solid rock.

Ribao took readings. The lava pool was three hundred meters below the top of the rim of Von Kármán, which provided a dense shield of basalt rock against cosmic rays and high-energy particles. The line of sight for his laser camera was one hundred meters. The ceiling of the cavern was twelve meters; the width of the entry was thirty meters. A perfect opening for lander/rovers and pickup-sized vehicles moving processed regolith. Ribao advanced one hundred meters before he encountered a ledge of about a meter high. This ledge was undoubtedly the result of some ancient vulcanism or earthquake. The ledge extended the width of the cavern. He decided to disembark from the rover/lander to continue his exploration of the site.

The astronaut checked his suit fittings and pumped the filtered cab atmosphere into a tank for storage. When atmospheric pressure was vacuum, Ribao vented what little air remained.

As an afterthought, the astronaut returned to the entrance and parked. He detached a Geiger counter which he affixed to the instrument panel.

He released the pressure seals on the pilot's door and jumped the short step to the ground, making sure that his tank of compressed air cleared the chassis of the lander. Ribao set the Geiger counter as close to the edge of the entrance as possible. Then he set out on foot to explore and chart the interior of the cave, not willing to risk the rover getting stuck in lunar dust or loose regolith. He checked his radio com. As he expected, the signal strength was zero bars. If anything happened in the confined area of the lava pool, Ribao could reach neither Mission Control nor his fellow astronauts working on the solar array. He was on his own.

Ribao had enough air to spend up to an hour charting the dimensions and condition of the empty lava pool. He would use his time wisely; his life may depend on judicious use of his air supply.

Lava pool, rim of the Von Kármán crater
March 25   1125 Hours Sol

Ribao paced the perimeter of the empty lava pool. He concluded that the space was adequate to shelter habitat/operations of the *Zhong Guo* colony for present requirements. The astronaut returned to retrieve the Geiger counter and fly back to base camp. As he approached the Geiger counter, readings on the device began to register strong solar activity, slowly at first, then spiking more rapidly. The physician retreated quickly deeper into the shelter of the basalt walls and ceiling of the lava tube.

Kourou Space Center, French Guiana
1125 Hours Sol

No previous manned expedition into space had ever met such an assault from the Sun. Unprotected by the atmosphere of the Earth, astronauts out in open space or on the Moon's surface were at risk of exposure to cosmic rays, X rays, gamma rays and high-energy particles. The deadly radiation came regularly in low amounts, or in intense concentration in the event of solar flares.

Mission control technician Ernest Simmons reported the solar flare after the computer had projected the direction that the intense burst of radiation would take. Travelling at the speed of light, the projection from the sun's corona would strike the Moon's far side and the Trojan asteroid shielding *Lunar Gateway* at L4 in five minutes.

"Attention. This is Mission Control. General Alert. Dangerous solar radiation is projected for your neighborhood by 1133 Hours Lunar. Any out-of-station activities must be conducted behind the shield of the Trojan asteroid. The duration of the unusual solar activity is unknown."

Von Kármán crater
Solar Array work site

March 25   1126 Hours Sol

The emergency message from Jiuguan Aerospace Center in Beijing was relayed by Queqiao – a communications relay satellite located at Lagrange point two. The commander's radio com flashed brightly on his helmet.

"Commander Huangdi – this is Jiuguan Aerospace. Acknowledge at once."

"This is Commander Huangdi."

"I provide emergency notice. The Aerospace Center has detected sunspot activity likely to result in a solar flare containing potentially lethal radiation. We have calculated those emissions to strike your general area as early as 1133 Hours Lunar. The duration of this solar activity is uncertain. What is your specific location?"

"Engineer Qinjaing and I are positioned on the assembly flyer at the highest level of solar collectors above the crater's rim. Qinjaing is tethered presently at the furthest extension of the top scaffolding. He will require a few minutes to disconnect."

"Dr. Ribao does not respond. Do you know his location?"

"He is using the lander/rover to explore the far end of Von Kármán crater. He is in the cliffside, out of radio contact."

"Dr. Ribao will have to fend for himself. Your orders are to find substantial shelter such as you can find immediately. Hopefully, the shelter you find will be adequate. When the radiation from the solar flare is over, Jiuguan will broadcast a continuous all-clear notification to you and Dr. Ribao."

"Affirmative," the Commander replied. "I will go now to retrieve Qinjaing with the assembly flyer."

Commander Huangdi quickly gathered what tools were strapped to the worksite. By the time he picked up Qinjaing, the display on the Geiger counter attached to the flyer was spiking wildly.

The Engineer pointed at the Geiger counter.

"Solar flare," explained the Commander. "I will try to find an outcropping of rock, a ledge where we can find some shelter from the radiation."

Perched on the flyer, the Chinese astronauts began a desperate search. Ten minutes under the lethal assault, Huangdi found a little cove on the rim's cliffside where the warning spike of radiation was abated. The two space-suited men huddled closely against the rock ledge, hoping for the best but fearing the worst.

Lava pool, rim of the Von Kármán crater
March 25   1152 Hours Sol

Time passes slowly for the astronaut waiting deep in a frigid dark lunar cavern for a solar flare to end. Twenty minutes of compressed air remained in his tank when a trip to the Geiger counter at the cavern entrance confirmed that the spurt of intense solar radiation was over.

Ribao's primary concern was for his crewmates Commander Huangdi and Engineer Qinjaing. Each of these men had obviously been irradiated while in a most vulnerable position – out in the open space - assembling the solar array with only the negligible shield of a space suit for protection against exposure to a lethal dose of high-energy radiation.

The mission physician made suitable time on his return to habitat module. He received a radio communication in route to the current habitat module.

"Ribao – did you evade the solar flare?"

"Yes. I was deep inside a basalt lava tube on the opposite side of the rim from the solar array when the heavy radiation arrived. How did you fare?"

"Engineer Qinjaing and I were working on the highest level of the scaffolding when I received notice from the Aerospace Center that a solar flare alert was received. I am moving at full thrust back to the habitat module."

"Qinjaing is in poor shape. He reported that he was weak and nauseous. I had to tie him to the assembly flyer to make sure that he did not fall off on the way to the habitat module. His face mask is smeared with vomit. He may be aspirated; he was coughing too much to speak."

"Was Engineer Qinjaing's exposure to radiation greater than your exposure?"

"Probably. Initially I received some shelter as I was working within the base of the scaffolding while Qinjaing was assembling solar collectors at the furthest reach of the scaffolding."

"Enough for now," said Dr. Ribao. "I will prepare treatment for maximum radiation sickness for you both. I have arrived at the habitat module. I will open the airlock for you. You need not wait for the air filter."

Von Kármán crater
Habitat/operations module
March 25   1235 Hours Sol

Dr. Ribao defrosted blood plasma from the freezer and prepared other steps for the emergency treatment of the irradiated crewmen. He lifted Qinjaing's bed bunk to the utility table and set out medication.

When Huangdi arrived at the module, Ribao assisted in carrying the limp body of the Engineer from the air lock into the module. The crewman's air tank was quickly removed. The physician suctioned Qinjaing's nose and throat before removing his space suit. Qinjaing was not breathing.

"Was Qinjaing breathing when you entered the airlock?" Ribao asked.

"The filters in the airlock are noisy. I do not think so."

"Would you say that was seven or eight minutes ago?"

"At least," replied the Commander.

Ribao made a mental calculation without comment. "Help me remove the space suit. I will suction his trachea and attempt to shock his heart."

Those efforts were unfruitful. After ten minutes, Ribao stopped the defibrillator. "The time of death was approximately 1235 Hours Lunar. We must record for the digital journal." The Commander and the Physician dutifully confirmed the death with voice recordings.

The immediate priority was to save the life of the Commander, if possible.

Ribao took a blood sample from Huangdi. The reading was 8.9 sieverts, an equivalent dose of exposure with a weighting factor for X rays and gamma rays. This factor gave Dr. Ribao an indication of how much the radiation had damaged living tissue compared with an equal dose of gamma rays or X rays. A dose of alpha particles causes about twenty times as much damage as the same dose of X rays. Doses above 0.70 sievert damage the parts of the body that produce blood cells. Death can result from infections and hemorrhaging within a few weeks, depending on the dose.

The reading from the blood sample taken from Huangdi was a death sentence, of which the Commander was aware. Nonetheless, Dr. Ribao began treatment with as much blood plasma as his body would tolerate. Anti-infection medication, anti-coagulants, and opioids with as much water as possible rounded out what treatment that the physician could perform at *Zhong Guo*.

The Commander took the unwelcome news in stride. "I will need your assistance. The microwave transmitter requires only an hour or so to connect and calibrate. We can complete this work and conduct a system check under my command. I want to complete the mission before I am too sick and weak to continue. My wife and son will be proud of a hero of the People's Republic of China."

Von Kármán Crater
Zhong Guo Solar Array
March 28

Dr. Ribao's engineering skills were challenged as he helped Commander Huangdi complete the assembly of the solar array and the connection required for the microwave transmitter. During the system check, Ribao recognized the change in Huangdi's voice and the short, deliberate movements which demonstrated his weakened condition. The commander barely survived the forty-kilometer flight on the assembler back to the habitat module. Huangdi needed assistance to his bunk. The Commander lay weakly on his bed roll, never again to arise on his own power.

Datang Utilities, the Chinese giant electric power company, conducted a secret test of *Zhong Guo* solar energy transmission. The reasons for secrecy were equal parts state paranoia and to prepare for a world which would have a diminished appetite for fossil fuels. The Chinese planned to dump over one trillion dollars' worth of holdings in the Athabasca tar sands alone. A suitable buyer – the Saudis – was currently being groomed. When the importance of the continuous one hundred mega-watt microwave transmission registered on world markets, the value of fossil fuels would take a further beating on world markets.

The Chinese had no great admiration for the Saudi royal family.

CHAPTER 34 – **Certified Sterility**

The Kingdom of Air

April 4

Telepathic summons for both Mephisto and Yuan Li were issued by Lucifer, Lord of the Air. Fear of the consequences of delay prompted the immediate response of each demonic legionnaire.

"In the digital age, the petrochemical-transportation (PTC) dollar is the prevalent commodity of exchange. The PCT dollar is under assault. The PCT dollar is the commodity which we can devalue and displace to the utter disconsolation of the world."

"How so?" asked Mephisto.

"A combination of attacks can shake the foundations of the PCT dollar. The first attack comes from the active promotion of a currency based on renewable energy credits managed under your able supervision."

Mephisto nodded. Normally he would have objected to yet another responsibility, but the demon was already working on this very project with *the Rugby Player*, one of the Council of Twelve's multi-billionaires.

"Registered owners of a kilowatt-hour credit will possess a digital account which could appreciate with credited money market interest, but not depreciate. Deposits will be in credits from transfers of kilowatt-hours of energy or acceptable currencies at current exchange rates. Due to balance of trade deficits and the poor state of the economies of many nations, many will choose electronic kilowatt credits from *Lunar Gateway* or *Zhong Guo* as a safe haven even over their own national currency."

"Is there a presumption of competition of microwave energy suppliers?" asked Yuan Li.

"Of course. There is a decided advantage to space-based solar collectors because solar energy collected in space receives 100% of the solar constant, whereas solar collectors on Earth only receive 40% of the sun's energy due to the atmospheric filter," Lucifer advised. "The efficiency of Earth-based solar collectors is further limited by the planet's day-time night-time rotation. Nonetheless, contributors to the electric power grid will include individuals selling household excess as well as large companies who construct large arrays of collectors."

"I understand that the Chinese have finally decided to sell their operations in the Athabasca tar sands, and that you have groomed a suitable buyer." Lucifer addressed his question to Yuan Li.

Yuan Li's demonic activities stretched millennia in the Orient. As the legionnaire controlling the devices of man in that area of the globe, he and his top deputies had a hand in every illicit transaction. "Yes," Yuan Li affirmed. "The Saudis are excited about acquiring tar sands leases and the existing operating companies in Canada. They are fools."

"You must expedite that transaction," Lucifer ordered. "I require the transfer of the tar sands properties by mid-May."

Mephisto understood the importance of the timing of that transaction and winked at Lucifer. The student was rewarded by a smile from his master.

Then, speaking to both legionnaires, Lucifer further instructed those public announcements be made of the recent deliveries of microwave energy from *Lunar Gateway* and *Zhong Guo* to electric power companies on Earth, also by mid-May. *The renewable energy source – the Sun – provided continuous power at one-third the cost per kilowatt hour as fossil fuels!* The timing of the public announcements would be important to a miraculous prediction that Mephisto would soon instruct Pope Luis I to deliver by world address.

"There exists yet another opportunity to demonstrate to the celestial Council of Worlds *the arbitrary nature of God's rule.* God permits the development of interrelated businesses – transportation and petrochemical, to name a couple – and allows the companies to operate and consume vast resources for the past 150 years – to poison the earth's environment to the tipping point of irreversible destruction! The obvious solution is stop using fossil fuels, but most of man's planned economies center around the use of these fuels. The transfer of the current economy to an economy based on renewably sources of energy will be very disruptive – hundreds of millions of people would lose their jobs. A global depression will occur with the attendant misery of homelessness and starvation. The actual cost of a transition from fossil fuels to renewable energy would be difficult to calculate, certainly trillions and trillions of dollars. So thusly, in God's plan, *mankind is damned if he does and damned if he does not.*" Lucifer appeared smug, and content with his conclusion. Lucifer provided specific assignments to his top legionnaires.

CRUDE OIL FUTURES COLLAPSE

Baron Financial (May 29). Following the announcements from the Kourou Spaceport and the Jiuguan Space Center that their outposts in space have tapped abundant solar energy available at a fraction of the cost of fossil fuels, international commodity markets reacted sharply. Amid slumping demand, the price of light crude oil futures fell to $25 per barrel before stabilizing. *Lunar Gateway*, a corporate contractor leasing the former International Space Station, reported that by month's end, their *Terawatt Project at L4* will sell solar power generating copious amounts of electricity to utility companies. According to authorized sources, a marketing arrangement featuring a currency of kilowatt credits was made with Elon Musk, the billionaire entrepreneur. Musk is the majority shareholder in SpaceX, which figures prominently in Artemis and Lunar Gateway . . .

Kingdom of the Air

May 30

A telepathic doorway opened in each of the minds of Yuan Li, Michael Popov and Mephisto. A golden clawed hand appeared at the entrance of the doorway, beckoning the demonic legionnaires. Without hesitation the demons complied, knowing that any delay would result in Lucifer's use of the dreaded tractor beam, from which resistance was futile.

The occasion was to witness from the netherworld the commission of Juan Romero as an operative chief of terrorist activities. After Johnny

267

Chen outlined Romero's responsibilities and confirmed his appointment, the psychic portal to Earth closed.

Instantly thereafter Lucifer's most effective agents found themselves seated in a richly- attired conference room. With J.S. Bach's *Fugue in D Minor* playing in sense surround, Lucifer made his dramatic appearance. Lucifer wore a three-piece suit with matching grey-and-white goatee.

Lucifer set his intense gaze on each of his legionnaires in turn. After a minute of silence, which was intended to induce contemplation, he posed a question. "What shall we take away from this object lesson?"

The rich baritone voice penetrated the demons' consciousness. They knew that the preferred answer would be forthcoming, so none of these fallen angels dared to respond.

Lucifer strode around the conference table. "My just accusations against our Adversary are known throughout the universe wherein our struggles are daily theatre for all created beings. These creatures recoil in horror from the events unfolding in what we know are preliminary to Earth's end-time destruction. Like the conflict between France and Prussia, the *Kingdom of the Air* faces a more powerful Opponent. Like Prussia, the key factor may be the sympathy and neutrality of some heavenly angels and created beings on other planets. Our struggle is primarily a spiritual battle between powers and principalities. We win when we gain entrance into the hearts and minds of those on Earth and on the Outer Worlds who profess allegiance to the Kingdom of Heaven. Our battle for hearts and minds takes place on many fronts. Destabilizing economies and inter-national civil order are important objectives. The United States is perceived by the world as the bulwark of democracy – a land of equal opportunity where justice prevails. When that democracy falters, a beacon of light that illuminates the world is dimmed. A combination of threats brought on by global climate change, terrorism and religious intolerance will extinguish that beacon of light."

"You have witnessed the commissioning of Juan Romero, a/k/a Zulfikar, who brought nuclear terrorism to the United States. He is one of many malcontents that you will employ in destabilizing the U.S. civil society."

"Of even greater significance is the United States' role in leading international religious intolerance. *As you are aware, the majority of churches in America are Protestant in name only.* **Gone is the significance of doctrinal teachings that differentiated Protestantism from Romanism. The evangelical and Pentecostal churches will be particularly useful to the False Prophet, the role that is America's destiny.**"

CHAPTER 35 – Certified Sterility

STERILITY INCREASES

Geneva, Associated Press (May 31)

The World Health Organization reported that sterility among young men markedly increased as demonstrated in tests conducted in sixty world metropolises. Men in countries in Asia and Africa are particularly affected; sterility is sharply rising among young men in South America. Medical research links the increased incidence of sterility to the current Gamma viral pandemic.

Writing in the *Journal of Virology*, Dr. Kwan Tseung, chairman of the Central Hospital Authority in Beijing, and Chou Xin, PhD, linked the Gamma H_3N_7 virus as hosting a variant human papilloma virus (HPV) as the direct cause of the sterility. The study indicated that in China as many as one-quarter of the male population who survived the viral pandemic may be impotent as a result of infection with the variant HPV.

Although new cases of the virus in China have declined, the World Health Organization continued to recommend a travel advisory alert for visitors to China.

Encrypted communique

June 1

the Cargo Ship: "As the Gamma viral pandemic winds down in much of the world, I am increasingly concerned about our exposure and liability.

the Hammer: "Outside of the Council of Twelve, I have spoken with no one regarding the virus. Thus far, Interpol has not come calling. How are you faring?"

the Cargo Ship: "When international trade suffers and the market for crude oil stumbles, business for the shipping industry badly falters. Industry-wide layoffs are forecast. I am doing everything that I can think of to prepare for the New World Order. At this point, I am still reluctant to even take subordinates into my confidence. I did not expect that our Gamma virus would be so effective or that our venture at *Lunar Gateway* would produce so spectacularly."

the Hammer: "We all underestimated the genius of our director *Q*. But the short-term outcome is devastating. In my construction and tool businesses, long-standing contracts have been curtailed; customer inventories of our products have been reduced. My companies have laid off ten thousand workers and look to further reductions in workforce as automation increases in manufacturing."

"Credit is tightening; cash is king. Even national currencies are suspect; I intend to deal in our new international currency – Kilowatt-hour Credits. I expect that our currency will quickly gain traction and even appreciate. Consider that we have established a pyramid scheme in which we are the originators; the significant difference, however, is that the bank that we have established in Kourou, French Guinea, backs our currency with credits redeemable in immediate kilowatt-hours. Our marketing director, Jeff Bezos, will direct billions of internet transactions through our new currency. You can bet that exchange rates will be favorable for members of the Council of Twelve."

the Cargo Ship: "In the aftermath of the Gamma pandemic, what opportunities do you see for business or investment?"

the Hammer: "That is for me to know and for you to find out. You employ some bright people; set them to work on business activities that can utilize cheap electricity. If I were you, I would not open any fertility clinics or buy stock in Gerber's Baby Foods."

Kowloon, China

June 1

Dr. Chou Xin and Dr. Emily Chung recognized the occasion of widespread infertility as a once-in-a- lifetime opportunity to open a national chain of fertility clinics. The eminent physicians, widely credited with establishing protocols effective against the viral pandemic, resigned from their positions at Kowloon Hospital Authority. They obtained financial backing and within a month, established a successful franchise operation. Tens of thousands of fertile sperm donors were identified and characterized for features. The physicians created and organized a sperm bank. Their director of operations touted the guarantees of their procedures to produce healthy fertile male babies.

Fees, however, were beyond the reach of many Chinese. The demand for infant male orphans became insatiable. Orphanages had waiting lists of over 2 million eligible Chinese men who are sterile.

CHAPTER 36 – **Military on the Moon**

the Von Kármán crater
June 3

Dr. Qianqian Ribao, physician and botanist for Cheng'e 6&7 missions, provided what medical treatment for Commander Huangdi as was available. As Huangdi's condition deteriorated due to lethal exposure to cosmic and gamma radiation, Ribao assumed all the responsibilities of command. In the week prior to his death, Huangdi was comatose.

Now Ribao was alone, one man on a vast expanse of land the size of the continents of Africa and Australia combined. Between radio contacts with Jiuguan Mission Control, Ribao tended the rice and soybeans in the hydroponics module. The Geiger counter was kept at hand's reach.

Dr. Ribao was rationing food and supplies of compressed air were short when the spacecraft Cheng'e 8 notified Ribao of its imminent arrival. "*Zhong Guo* Base Camp, this is Colonel Sun yet-Sen, commander of Cheng'e 8 spacecraft. Come in, Dr. Ribao."

"This is Ribao. What is your estimated time of arrival (ETA)?"

"In one hour, starship Cheng'e 8 will jettison from lunar orbit the equipment and supplies for our mission. Robotic dirigibles will guide these materials to soft landings. Thereafter our lander will separate and auto-brake, using thrusters to reduce our speed before landing. I bring a crew of six."

The landing was without incident. A large cloud of lunar dust blew off the Cheng'e landing pad. Upon arrival, the military crew of six men suited for the almost-airless conditions, departed their spacecraft, and quickly went to work. At the temporary Base Camp in the midst of the Von Kármán crater, a radiation shelter was dug in the regolith, deeply enough to be effective against the punishing rays of the sun.

Most of the equipment and supplies had been dropped at the new site for *Zhong Guo*. The permanent site for the Chinese base of operations was the vacant lava pool, a large cavern discovered by Ribao on the rim of the crater. Although this site was located on the opposite side from the one hundred mega-watt solar collector array and seventy kilometers distant, this arrangement was deemed satisfactory by officials of the Chinese Lunar Exploration Program (CLEP).

A module was inflated in the cavern for temporary accommodation. After digging a trench and laying PVC pipe for the collection of organic waste, the military construction crew built a wastewater/ collection/ recycling facility suitable for the use by as many as eighty people. A wire harness for electricity was laid out to power light and heat for the habitat. Then the crew sealed and insulated a portion of the cavern for habitation in modules placed in the lava pool cavern. The cavern outside the insulated modules, far from the penetrating radiation of the sun, was one of the coldest spots in the universe.

With the lessons from dealing with lunar dust in mind, airlocks were constructed with adequate filtration and venting of fine particles. The fine particles were usually crystalline silicates, often suspended in layers above the lunar surface due to the Moon's light gravity. The attention paid by mission planners to the resolution of a fundamental problem was gratifying to Ribao. Ignoring this serious matter could bring Chinese operations on the Moon halting as fatal silicosis developed in the lungs of the lunar astronauts.

In the course of their duties, Dr. Ribao and Commander Sun yet-Sen often conferred about health and safety risks. When tasks related to the habitat for workers drew to completion, Ribao posed a question to the Commander. "Is the primary objective of your mission the construction of another one-hundred mega-watt solar array, or do you intend to start mining regolith for the extraction of ice water, tritium, lithium, and helium-3?"

"Although those objectives are priorities for *Zhong Guo*, my orders are first and foremost to build a military outpost prepared for defense of our colony," said Col. Sun yet-Sen. "Barracks and support facilities for soldiers will be constructed. Furthermore, I plan to harden this site against missile strikes. I will lay superconductive rails for cannon emplacement. *This site will become a base from which to operate against those who challenge the Chinese claim to lunar territories.*"

Lunar Gateway
June 10

When a cancellation of a reservation aboard the *Virgin Galactic* spaceship opened a spot for another space tourist, David Rothschild seized the opportunity to see firsthand the expanded modules of the *Lunar Gateway*. The flight may prove to be a once-in-a-lifetime chance for space travel and to tour the ISS. The time had come to interview the leading candidate for the permanent position of on-site Manager. Captain Dustin

Garrett had finished his six-month rotation and was scheduled to return to Earth within the month.

Figure 37. techcrunch.com

Once the space flight of the *Virgin Galactic* was underway, Capt. Garrett was notified of the impending visit of an important corporate sponsor. Garrett's curiosity was aroused. The flight arrangements of the VIP were scheduled and paid for by the European Space Agency (ESA); an appointment with Capt. Garrett was made for the unnamed VIP by Morgan Trussler of G45. After the elliptical flight to Lagrange point four, the *Virgin Galactic* docked at the ISS. During the tour of the ISS and Lunar Gateway modules, Garrett noticed that one of the guests had an ESA label on his spacesuit where the name was typically displayed. Garrett made the connection and presumed that he had found his mysterious guest-passenger. At an opportune moment of relative privacy, the captain approached the ESA visitor.

"Will you follow me to the Observation Center? The Center is currently unoccupied and may be appropriate for our discussion."

The visitor nodded wordlessly and accompanied the captain through a series of modules and hallways which marked the growth of the ISS/*Lunar Gateway*. They made their way to the outer wall of the station which provided a real-time view of the large Moon with the smaller blue Earth in the background. The spectacular view was the result of an outside camera which twisted and rotated on its mounting to present the best visual perspective for visitors. Those views were presented on a grid of monitors that stretched from floor to ceiling and from wall to wall. As far as the visitors were *co*ncerned, the views of the Earth and the Moon were spectacular, as good as they would get, short of a suited spacewalk. The Observation Center was a popular spot for visitors and crew alike.

"Are you the unnamed passenger?" Garrett asked.

"Yes. For security reasons, I am called *Q*. I am one of your Directors. You have come to our attention as a suitable candidate for full-time Manager of the *Lunar Gateway* facility. There is no application process. Your abilities and achievements speak for themselves."

"Thank you," Garrett said simply. There had been many opportunities for personal publicity which the Captain had routinely declined, mostly due to his busy schedule. Also, he had concern about keeping his position. He loved his job. Garrett drove himself intensely to accomplish the demands of his position. Other than the cooperation that he enjoyed from mission planners at Kourou Space Center, there was no boss on-site looking over his shoulder. He had long suspected that the European Space Agency was not his real employer. For a crew of eighteen, there was no director of personnel nor any unnecessary meetings. As his six-month rotation drew to a close, Garrett had no specific plans for his next employment.

The six-month rotation was dictated for reasons of optimal health. Mostly due to the near-weightlessness conditions inside the space station, the long-term prognosis for medical problems, such as loss of bone density, was likely. A recent discussion with mission planners revealed a plan to separate modules of the *Lunar Gateway* and attach thrusters that would give them a perpetual spin. Centrifugal force could replicate gravity.

Figure 38. https://www.bbc.com/future/article/20130121-worth-the-weight

The old and fragile International Space Station had barely survived the trip from low-earth orbit to Lagrange point four. Subjecting the ISS to centrifugal force would tear it apart. Garrett found himself taking a very personal interest in the plan to separate from the old ISS, if for no other reason than to enhance his own longevity as a resident of the *Lunar Gateway*.

Garrett was middle-aged; he had no family or real connections back on Earth. Garrett had some opportunities for employment back on earth, but his keenest interest lay with extending his tour in space for another six-month rotation. Garrett derived some hope in that regard by the relocation of the ISS/*Gateway* behind the Trojan asteroid for the purpose of shelter from lethal radiation. Garrett saw great promise for the *Lunar Gateway*. Whatever the position that *Q* was offering, the visitor had Garrett's full attention.

"Give me an idea of what the future holds for *Lunar Gateway*," Garrett requested.

Q admired Garrett's sense of priorities. The man was motivated primarily by a sense of purpose, not compensation and benefits. "*Lunar Gateway* will be a permanent settlement in space, providing energy to mankind and fuel to visiting spacecraft. You will build a second large array of solar collectors for *Terawatt Project at L4*. You will exploit the Trojan asteroid for water and mineral resources and head the exploration team to the Moon's South Polar region. Most importantly, you will prepare for the military defense of *Lunar Gateway*."

Dustin Garrett's eyes glistened. "I am the man for the job.'

The End of *The Conspirators*

ENDNOTES

[1] "RUSSIA WILL WITHDRAW FROM ISS'. *The New York Times* (August 7, 2022).

[2] Ibid. Source: https://nl.nytimes.com/f/newsletter/2022-08-07.

[3] https://www.NASA.org (<u>Saturn V</u>, <u>Space Shuttle</u>, <u>Space Launch System</u>), www.SpaceX.org (<u>Falcon 9</u>, <u>Starship</u>)

[4] Blanco, J., *The Clear Word* (2006), Hagerstown, MD, 2 Peter 3:10, p.1286.

[5] Blanco, 1 Peter 1:13, p.1279.

[6] Source: https://www.nytimes.com/spotlight/joe-biden

[7] Ibid.

[8] Source: https://www.defense.gov › Releases › Release › Article/3083102/fact-sheet-on-security-assistance-to-ukraine

[9] Ibid.

[10] Andrews Study Bible, NKJV (2010), Andrews University Press, Psalms 127:1, p.774.

[11] Madelyn Murray O'Hair, *What on Earth is an Atheist?* (New York), Arno Publishing, p.41.

[12] E.G. White, *Steps to Christ*, Harvesttime Publishing (2011), Altamont, TN, p.14.

[13] https://www.marinebio.org › creatures › zooplankton

[14] PLA Jou*rnal of National Defense. https://www.janes.com>posture-for-WWIII*.

[15] *PLA Journal of National Defense. https://www.janes.com>space-based-weaponry*.

[16] "*Azolla/Nostoc – Superorganism Substitute for Fertilizer*," *News & Observer*, Raleigh, NC, February 13, 2013, ppA7-A8. Article cites Francisco Carrapico, author of *Symbiosis and Stress*.

17 Ibid.

[18] Blanco, *The Clear Word*, (2006), Hagerstown, MD, Job 1:6 and Job 2:1-3, p.569.

[19] *Good News Bible*, American Bible Society, NY (1979), Romans 8:6-8

[20] Blanco, *The Clear Word* (2006), Hagerstown, MD, Revelation 12:12-13, p. 1307.

NEXT NOVEL

TOXIC MESSAGE

A plausible path of prophetic historicism

- PREVIEW OF SOON-TO-BE-PUBLISHED NOVEL -

Acknowledgements

There are numerous people to whom I am indebted. Portions of this novel was first published as ***Entry to Alliance Empire*** and ***Alliance Regime***. Those novels were my first books. I kept the theme and fixed many technical flaws. ***Pathway to Destiny*** is my third novel and ***The Conspirators*** is the fourth. ***Toxic Message*** is the fifth. I credit Tony Wood who provided a listening ear and constructive criticism night after night. I can never give enough thanks to my sisters Laurie Salmons, Cindy Reinhardt, Carolyn M. Moe, and my friend Alina Ancheta who provided support and encouragement. They were always willing to listen to my predictive ravings and my theories of how big problems will be resolved.

Preview of TOXIC MESSAGE

282

CHAPTER 1 - **Collateral Damage**

Airspace over Syria

September 28

The AWACS circled for over two hours over the territory between Aleppo and Damascus. AWACS, abbreviation of **Airborne Warning And Control System**, is a mobile, long-range radar surveillance and control center for air defense. The system, as developed by the U.S. Air Force, is mounted in a specially modified Boeing 707 aircraft.

Lt. Gen. Michaels had ordered special authorization for use of two AWACS and four crews for an extended surveillance of Syria. A drone attack was planned on a high-priority target. The mission was in its third day.

General Michaels was observing several activities remotely from his command post at Langley Air Force Base in Virginia. This mission was a live test of the capabilities of the new Defense Support System satellites, soon to be incorporated into Space Command's Orbital Laser Defense (SCOLD). His station monitored the AWAC, and the satellite assigned to the Syria area. The satellite had spotted the black armored Mercedes five days ago from its geosynchronous orbit. Normally the AWAC would order the release of the drone; this time the target would be acquired by the satellite, and the laser from the satellite guided a Hellfire missile from the drone to impact.

This action was an initial test for the capabilities of the DSS satellites. Other features would be tested in short order.

General Michaels did not concern himself with either the politics or the justice of the destruction of authorized targets. As a soldier, Michaels did not carry the burden for the ethics of 'pushbutton assassination.' Nor did he make the decision of which target to destroy. Those decisions were based on intelligence from the CIA. Therefore, General Michaels was absolved of the responsibility of guiding a missile attack on the wrong black Mercedes. It was certainly not the first time that a USAF drone dispatched a target in error, nor perpetrated 'collateral damage.' The incident would not be the last of its kind.

***

Aleppo, Syria   October 7 12PM

The Muslim funeral procession had left the mosque minutes earlier. A mostly empty casket was carried on the shoulders of friends and family. General Salim Idris deeply mourned the death of his son-in-law, Rashid Suliemon. Suliemon was a captain and a trusted adjutant in the Free Syria Army. The Syrian freedom fighters would miss his bravery and leadership skills in combat. Desi, Salim's daughter, could not be consoled. Rashid's death left her a widow, and Salim's grandsons were without their doting father.

A diplomat from the US State Department issued a statement of regret, as the intended target had been an Al-Qaida commander. Yet from an operational standpoint, the United States hardly distinguished the different combatants who fought the dictator Bashar Assad. Since Al-Qaida also fought against Assad, increasingly U.S. policymakers viewed the Free Syrian Army as tainted by association with Al-Qaida. Nothing could be further from the truth. There was no planning or coordination between the two anti-Assad factions. Often there were running gun battles between the FSA and Al-Qaida fighters.

The civil war was developing into a sectarian strife, with battle lines of Hezbollah against Al-Qaida, and Alawite Shiites versus radical Sunni Moslems (ISIS). The Syrian Kurds were carving out their own section of Syria. For the Coptic Christians, Assad was the secular devil they knew and was preferred over a potential Islamic state which would result from a rebel victory. Neither Hezbollah nor ISIS made provisions for Christians in the new regime.

The actions and inactions of the United States were infuriating; as an ally against Assad, the U.S. was worse than unreliable. At first the U.S. promised to supply arms, and to enforce a "no fly" zone which would effectively ground Syrian jet fighters and attack helicopters. The U.S. had also promised economic boycotts and financial assistance. Yet none of this aid from the U.S. ever materialized. Now it appeared that the United States favored the status quo, which was catastrophic. Since 2012, the civil war had claimed over 500,000 dead and twice that many with lifetime injury. Over eight million refugees were languishing in camps in Jordan, Iraq, Lebanon, and Turkey. Many survivors were starving, hidden away in war-torn ghettos. American policy -action and inaction - was killing Syria. Most Syrians believed that the U.S. planned to partition the country.

As he watched his son-in-law's funeral procession slowly weave through the narrow streets of Aleppo, Salim was coldly calculating his revenge for Rashid's death and the needless deaths of so many Syrians. America was to blame. It would be justified, he thought, if there was American blood spilled for Syrian blood.

CHAPTER 2 – **Russian Arms Dealer**

Johannesburg, South Africa

October 14

A lifetime of experience made Anton Yukolopov a cautious man. This was an essential trait for survival in the business of selling weapons. His longtime boss and mentor Victur Bouk had once told Anton that there were old arms dealers and there were bold arms dealers, but there were no old, bold arms dealers. Victur was now serving a twenty-year prison sentence. Five years ago, Victur broke one of his own cardinal rules, and a "sting" operation conducted by a military contractor seized him. Victur was imprisoned in the Netherlands. With the approval of the Russian president, Anton inherited a lucrative arms trading concession.

The business was immensely profitable. There was a continuous demand for Anton's weapons. His inventory of arms was extensive. What weapons Anton did not keep in stock; he could acquire – for the right price. Business was good, with no shortage of men who wanted to kill other men or keep a nation of malcontents under control.

For the past five years, Anton's most lucrative business was with factions of the anti-Assad Syrian resistance. Arms shipments were made regularly with secure payments arranged through the Saudi Crown Prince and the Sunni Moslem emirates of Qatar, Oman, Bahrain, and Dubai. Anton's profits were tens of millions each year from the conflict in Syria. Ten percent of gross sales of weapons to Assad's opponents was set aside as personal income for Russia's president.

Since Russia's military propped up Assad's regime, Anton could only guess why the Russian president permitted weapons sales to the Free

Syrian Army and their allies. Beyond fattening his bank account, what was the president's strategy? Long-term destabilization of Syria, with the misery of ten million refugees visited upon neighboring Arab countries? Belittling America's image abroad as a dependable ally in a shooting war? Further establishment of Russia as the world's premier vendor of military arms and munitions? Setting the stage in the near future for a proxy war in which the United States supported Israel while Russia backed a host of regimes eager for the annihilation of Israel? All these scenarios could fit.

Anton was concerned that much of that business was in jeopardy. This morning Anton had received a message from his best customer – the Free Syrian Army, based in Turkey. The message was sent in proper security protocol, encrypted, to Anton's scrambled IP address. The Syrian customer wanted to purchase a miniature nuclear device, and the customer had reliable information to believe that one was available from the Pakistanis.

Anton had acknowledged receipt of the inquiry – that response had been automatic. What should he do? Victur Bouk, his old boss, had another old Russian adage that was applicable: 'Pigs get fat, but hogs get slaughtered.' Victur would have rejected the request immediately, without hesitation. That was the wisest course of action. Anton began to send such a message to the Syrian, and then, he reconsidered.

The Russian pondered his position further. As a result of the chaos following the horrendous earthquake in Pakistan, Anton had been buying U.S.-made arms from officers in the Pakistani military for pennies on the dollar. The weapons, ammo, and other military equipment that he received were often in the same crates in which the US had shipped them. Anton was receiving this equipment by land, sea, and air. Anton's warehouses were full to overflowing. Business was booming, but cash flow was a problem, with the all the outgoing payments.

Anton learned from a reliable source that a Pakistani officer had 'sequestered' a miniature nuke for *safekeeping* immediately after the massive earthquake. The message from the Syrian said that he did not trust the Pakistanis and that he would prefer to deal with the Russian. There were several good reasons for the Syrian to distrust the Pakistanis. The first was religion – the Muslim sects were at war with each other. The second was that Pakistan was contaminated by their long-term association with the United States and particularly, the CIA. The whole prospect of 'dealing' a miniature nuke could be CIA entrapment.

Finally, Anton concluded that the mere availability of the weapon was a threat to his business. The Syrian had not indicated the purpose for the bomb. What was the intended victim? Damascus? Tel Aviv? Or . . . the United States?

The Russian knew well the blood feud against Assad, Hezbollah, and the Alawites. Certainly, most Arabs, and many Muslims, hated Israel. But detonating a nuclear weapon in Damascus or Tel Aviv? What would that accomplish? Bombing Damascus would destroy what the rebels were trying to save. The security into the port of Tel Aviv was legendary. Entry through port authority, by air or through checkpoints on land was extremely problematic. Certainly, the Syrian must intend to send this 'package' – the nuke - by cargo container to a U.S. port city. The U.S. vulnerability was widely known; only 5% of all sea-going containers were inspected by U.S. custom officials prior to entry at port.

Anton thought further. Yukolopov's arms trade was insulated in South Africa and the other countries from which he stored and shipped arms. Thanks to the handsome bribes paid to numerous government officials, and his relationship with the Russian president, Anton was almost untouchable. His business activities were tolerated by the host countries. Currently Anton was evading service of a subpoena from the United Nations' war crimes tribunal. Furthermore, Interpol had an ongoing investigation into Anton's arms trade. Although Interpol has extensive resources, as long as he remained in the good graces of the Russian President, Anton was untouchable.

But if there was a nuclear explosion in an American city, and Anton was implicated, there would be nowhere in the world that he could hide. Because Anton was the principal arms buyer from the Pakistani military since the earthquake, would the Americans assume that Anton had dealt the nuclear device? Undoubtedly.

Anton made up his mind. The arms dealer would find the weapon and purchase it. Then he would deal the weapon to the Americans in some fashion. What would the Americans pay for the information? What would they pay for the nuclear device? Mere knowledge of the suitcase-sized nuke by the Russian agent was sophisticated extortion – the nuke was in the hands of rogue mercenaries. Anton concluded that the Americans could not risk leaving the nuclear weapon in circulation among the Arab states.

Anton saw opportunities in purchasing the suitcase bomb. He presumed that he could deal the nuke to the Americans for a nifty profit while enhancing his value to the Russian President. Goodwill generated by the proper handling of the matter could be leveraged to ease burdensome economic sanctions.

What to do? Anton needed obtain a cash deposit from the Syrian buyers first, then find the Pakistani army officer who had the bomb.

All of Anton's plans were subject, of course, to the approval of the Russian president. The Russian president was Anton's business partner, yet he was the only man that Anton truly feared. Enemies of the president and often those people who annoyed the president died suddenly and anonymously. The methods of inflicting his wrath differed – garrote, poison, a silenced 9mm round to the head – which were equally effective.

Anton made certain that his accounting was current and accurate. Anton called the Russian President, exchanging peasantries. He made an appointment. Then he boarded his private jet. His destination was Moscow. Anton knew that the Russian President would greet Anton warmly and that he would determine a profitable and politically expedient way to dispose of a miniature nuclear device. Anton was certain that the Russian President would find some way to embarrass or injure the United States.

,

CHAPTER 3 - **Laying Groundwork**

In the supply line through which arms and ammunition were procured, Anton Yukolopov consolidated the network of agents among the Pakistani army officers. Some officers he rewarded by giving increased responsibilities with increased pay. Others, unreliable or potential trouble-makers, Yukolopov released. He stream-lined the purchase of military hardware under Col. Aziz Khan.

Khan, a professional soldier, recognized that the recent earthquake and the immensity of its destruction created many new realities. One reality was that Pakistan would be an economic basket case for years to come; the second reality was that Pakistan's military would be unable to defend its borders. At best, the military could provide emergency disaster services and help restore order. At worse, different units would ally with warlord chieftains, and Pakistan would disintegrate. Already the Pakistani army had abandoned their posts within the disputed province of Kashmir.

Kokopo had offered Khan a lucrative assignment – locate the miniaturized nuclear weapon as discretely as possible and make a reasonable offer. Khan heard of such a weapon and with proper military support was certain he could acquire the mini nuke. Yukolopov decided to dispatch a trustworthy courier with a cash payment of $500,000 U.S.

Three attack helicopters accompanied Khan when he met the Pakistani army officer who had the mini nuke. The officer wisely took the money and fled with his small cadre of soldiers.

On October 24, Yukolopov received a $4 million deposit into the bank of Caracas in Venezuela from the Saudi Crown Prince.

That same day, Yukolopov called the U.S. Central Intelligence Agency in Karachi, Pakistan. Yukolopov's reputation had preceded him.

"Station 27 – Porterfield here." The switchboard identified the origin of the call was Johannesburg, South Africa. When Allen Porterfield heard the heavily accented Russian speak, the CIA operative had a good idea with whom he was speaking. Flawlessly, Porterfield switched his speech from English to the Russian language. *"Dobriy den"* (Good afternoon). "Privet" (Привет). Porterfield was fluent in Russian.

The Russian did not waste time with small talk. "Due to the destruction of the Atomic Energy Commission facility in Peshawar, rogue elements of the Pakistani Army have acquired a small nuclear weapon. I cannot seize it without risking it detonation. Although the mini nuke is small by comparison with the typical payload of an Inter-Continental Ballistic Missile, the concussive blast is equivalent to two hundred tons of TNT."

"Thanks for bringing this information to me. Should I file this information for future reference?" Porterfield was certain that he would obtain every detail that Yukolopov was authorized to release.

"You would ignore or dawdle on this grave matter at significant risk to America or her allies. Even now, those who possess the mini nuke are seeking to strike a deal for its purchase."

"I realize that your options to act are limited. Do you have a proposal whereby we could seize this mini nuke or otherwise take it off the market?" Porterfield asked.

"As a subterfuge to gain the confidence of the rogue elements now in possession of the bomb, I will agree to broker the purchase to certain

Middle East factions which currently buy millions of dollars of arms, ammunition, and military vehicles."

"Assuming the CIA agrees to purchase the bomb and transport it safely out of Pakistan, how do we know that we haven't bought a chunk of scrap metal?"

"Why don't you send your own expert nuclear technician to ensure the bomb's authenticity? While your technician is examining the mini nuke, I suggest that he do whatever is possible to disarm it. I will make such an inspection a condition for its purchase. I will accomplish this for $3 Million." The Russian's proposal sounded quite reasonable when contrasted with the catastrophic damages and loss of life which was an alternative.

"Consider your proposal as accepted, subject to approval by the CIA Director Dan Clay."

"Do svidaniya." "до свидания."

"Do svidaniya." The international call terminated.

Yukolopov considered with satisfaction his past negotiations with the CIA. The deal Yukolopov wrangled on this occasion would have won the admiration of his old boss Victur Bouk. Bouk languished in a Russian prison for 15 years for undercutting the Russian President in an arms deal.

***

The CIA gave their courier $3 million for Yukolopov's part in the CIA plan. The CIA also promised to 'fix' Yukolopov's problems with Interpol and the United Nations. The CIA planned to seize the weapon, but not in Pakistan because the Pakistani government would be outraged. The

CIA clearly planned some political use for the mini nuke. Yukolopov did not care what use the CIA made of the mini nuke if the Russian president signed off on the sale. His role was finished when Col. Khan delivered the device in its lead enclosure to the port at Karachi. The Pakistani soldier made plans to deliver the bomb to the Syrians.

The End of the first three Chapters of *Coming Apocalypse,* written by W. R. Reinhardt. Illustrations (figures) in the form of photographs, drawings, diagrams, and graphs are expected to accompany text. Hardcover and eBook editions will be published. Cover art is designed by R L Salter, produced by Self-Pub Book Cover, printed by BookBaby, copyright and ISBN registrations are in possession.. All rights are reserved. *Coming Apocalypse* can be preordered from Barnes & Noble, Amazon, BookBaby, and Target in December 2022.

ANCIENT ALIEN DEMONS CONSPIRE WITH HUMAN HENCHMEN TO DEPOPULATE THE EARTH

In **The Conspirators**, by W.R. Reinhardt, Earth balances precariously at the tipping point of survival due to global climate changes. The likelihood looms of Mutual Assured Destruction in an exchange of Intercontinental Ballistic Missiles. In order to save Earth, invisible Aliens manipulate a powerful cadre of human Illuminati. The multibillionaires vow to reduce the planet's population by ninety percent within one generation through perpetual warfare, viral pandemics, and environmental destruction.

The Immortals invoke their plan to save Earth from complete environmental destruction and shed the cruel chains of Alien bondage from the humans. The Immortals seek allies among the humans who would be suitable citizens of a civilization of peaceful Immortals who would thrive in an Earth made new.

The Aliens also seek allies among the humans, offering inducements that include riches, fame, power, and a lifetime of pleasurable pursuits. As the Alien demons work their will into the psyche of their human allies, the humans become unwitting hosts, possessed by demons.

The author mixes literary techniques akin to John Bunyan and C.S. Lewis in a tale wherein demons are constrained by *Rules of Engagement*, and Lucifer teaches object lessons to his Legionnaires in holographic chambers and in live events. Holy angels enforce the *Rules* which prevent demons from wielding undue influence.

This sci-fi novel engages societal norms, religion, and philosophy in a contemporary satire. The events of the novel certainly summon current big-picture issues like global climate change, perpetual war, nuclear disarmament, viral pandemics, civil unrest, and human trafficking.

IN A DISTURBING YET ENTERTAINING FASHION, REINHARDT DEMONSTRATES THE ANALYSIS OF A DEEP THINKER WHO VOICES CONCERNS ABOUT CURRENT TRENDS WHICH DEFY SIMPLE SOLUTIONS.